NELSON VICmaths

VCE UNITS ① + ②

general mathematics 11

mastery workbook

Dirk Strasser

Nelson VICmaths General Mathematics 11 Mastery Workbook
1st Edition
Dirk Strasser
ISBN 9780170464086

Publisher: Dirk Strasser
Additional content created by: ansrsource
Project editor: Alan Stewart
Series cover design: Leigh Ashforth (Watershed Art & Design)
Series text design: Rina Gargano (Alba Design)
Series designer: Nikita Bansal
Production controller: Karen Young
Typeset by: MPS Limited

Any URLs contained in this publication were checked for currency during the production process. Note, however, that the publisher cannot vouch for the ongoing currency of URLs.

Acknowledgements
TI-Nspire: Images used with permission by Texas Instruments, Inc
Casio ClassPad: Shriro Australia Pty. Ltd.

© 2022 Dirk Strasser

Copyright Notice
This Work is copyright. No part of this Work may be reproduced, stored in a retrieval system, or transmitted in any form or by any means without prior written permission of the Publisher. Except as permitted under the *Copyright Act 1968,* for example any fair dealing for the purposes of private study, research, criticism or review, subject to certain limitations. These limitations include: Restricting the copying to a maximum of one chapter or 10% of this book, whichever is greater; providing an appropriate notice and warning with the copies of the Work disseminated; taking all reasonable steps to limit access to these copies to people authorised to receive these copies; ensuring you hold the appropriate Licences issued by the
Copyright Agency Limited ("CAL"), supply a remuneration notice to CAL and pay any required fees. For details of CAL licences and remuneration notices please contact CAL at Level 11, 66 Goulburn Street, Sydney NSW 2000,
Tel: (02) 9394 7600, Fax: (02) 9394 7601
Email: info@copyright.com.au
Website: www.copyright.com.au

For product information and technology assistance,
in Australia call **1300 790 853**;
in New Zealand call **0800 449 725**

For permission to use material from this text or product, please email
aust.permissions@cengage.com

ISBN 978 0 17 046408 6

Cengage Learning Australia
Level 7, 80 Dorcas Street
South Melbourne, Victoria Australia 3205

Cengage Learning New Zealand
Unit 4B Rosedale Office Park
331 Rosedale Road, Albany, North Shore 0632, NZ

For learning solutions, visit **cengage.com.au**

Printed in China by 1010 Printing International Limited.
1 2 3 4 5 6 7 26 25 24 23 22

Contents

1. Investigating and comparing data distributions — 1
Matched example 1	Deciding on the type of data	1
Matched example 2	Finding the median, mode and range	3
Matched example 3	Constructing a frequency table	4
Matched example 4	Constructing a grouped frequency table	5
Matched example 5	Reading a bar chart	6
Matched example 6	Reading and describing histograms	7
Matched example 7	Finding the five-number summary by hand	9
Matched example 8	Finding outliers	10
Matched example 9	Reading boxplots	11
Matched example 10	Using dot plots	12
Matched example 11	Using stem plots	14
Matched example 12	Working with back-to-back stem plots	16
Matched example 13	Interpreting back-to-back stem plots	17
Matched example 14	Working with parallel boxplots	18
Matched example 15	Calculating the mean	19
Matched example 16	Working with the mean and standard deviation from a display	20
Matched example 17	Working with the mean-standard deviation scale	21

2. Arithmetic sequences and financial recurrence relations — 22
Matched example 1	Finding sequences from rules	22
Matched example 2	Identifying sequence types	24
Matched example 3	Graphing arithmetic sequences	25
Matched example 4	Finding arithmetic sequences from recurrence relations	27
Matched example 5	Finding the nth value of an arithmetic sequence	28
Matched example 6	Using simple interest recurrence relations	29
Matched example 7	Using recursive computation for simple interest	31
Matched example 8	Using the simple interest general rule	32
Matched example 9	Using flat rate depreciation recurrence relations	33
Matched example 10	Finding the rate for flat rate depreciation	34
Matched example 11	Using the flat rate depreciation general rule	35
Matched example 12	Using unit cost depreciation recurrence relations	36
Matched example 13	Using the unit cost depreciation general rule	37

3. Geometric sequences and financial mathematics — 38
Matched example 1	Graphing geometric sequences	38
Matched example 2	Finding geometric sequences from recurrence relations	40
Matched example 3	Finding the nth value of a geometric sequence	41
Matched example 4	Comparing compound and simple interest	42
Matched example 5	Working with compounding periods	43
Matched example 6	Finding compound interest recurrence relations	44
Matched example 7	Using the compound interest rule	45
Matched example 8	Finding reducing balance depreciation recurrence relations	46
Matched example 9	Using reducing balance depreciation recurrence relations	47
Matched example 10	Using the reducing balance depreciation rule	48

Matched example 11	Calculating mark-ups and discounts	49
Matched example 12	Finding the percentage change r	50
Matched example 13	Finding the original price	51
Matched example 14	Working with GST	52
Matched example 15	Using the unitary method	53
Matched example 16	Using the unitary method for percentages	54
Matched example 17	Using the unitary method to make comparisons	55
Matched example 18	Finding single item inflation rates	56
Matched example 19	Working with inflation rates	57
Matched example 20	Calculating the purchasing power of money	58
Matched example 21	Understanding personal loans	59
Matched example 22	Comparing purchasing options	60

4 Linear functions, graphs, equations and models — 61

Matched example 1	Drawing linear graphs from tables of values	61
Matched example 2	Determining if a point lies on a line	62
Matched example 3	Finding the slope using $\frac{\text{rise}}{\text{run}}$	63
Matched example 4	Finding the slope from two points	65
Matched example 5	Finding the straight line equation from the slope and intercept	66
Matched example 6	Working with the constant rate of change and initial value	67
Matched example 7	Dealing with the domain of interpretation in real-life problems	69
Matched example 8	Modelling profit and loss	70
Matched example 9	Graphing linear relations in the form $Ax + By = C$ by hand	71
Matched example 10	Solving problems using simultaneous equations	72
Matched example 11	Interpreting line segment graphs	73
Matched example 12	Interpreting line segment distance-time graphs	74
Matched example 13	Interpreting step graphs	75

5 Matrices — 77

Matched example 1	Understanding the order of matrices	77
Matched example 2	Identifying types of matrices	79
Matched example 3	Adding, subtracting and multiplying matrices by a scalar	80
Matched example 4	Finding missing elements in matrix equations	81
Matched example 5	Working with matrices using addition, subtraction and scalar multiplication	82
Matched example 6	Multiplying matrices	83
Matched example 7	Finding the determinant and inverse of a 2×2 matrix	85
Matched example 8	Solving simultaneous equations with two unknowns using matrices	86
Matched example 9	Solving problems using inverse matrices	87
Matched example 10	Solving costing and pricing problems using matrices	88
Matched example 11	Working with communication matrices and diagrams	90
Matched example 12	Working with two-step communication	92
Matched example 13	Constructing transition matrices	93
Matched example 14	Interpreting transition matrices	94
Matched example 15	Finding the steady-state matrix	95

6 Relationships between numerical variables — 96

- Matched example 1 Identifying explanatory and response variables — 96
- Matched example 2 Interpreting scatterplots — 97
- Matched example 3 Scatterplots and association — 98
- Matched example 4 Exploring causation — 99
- Matched example 5 Rounding to decimal places versus significant figures — 100
- Matched example 6 Working with the line of good fit equation — 101
- Matched example 7 Interpreting a line of good fit equation — 103

7 Graphs and networks — 104

- Matched example 1 Identifying isomorphic graphs — 104
- Matched example 2 Identifying the features of graphs — 105
- Matched example 3 Finding adjacency matrices — 106
- Matched example 4 Verifying Euler's formula — 107
- Matched example 5 Using Euler's formula — 108
- Matched example 6 Identifying subgraphs — 109
- Matched example 7 Classifying walks shown on a graph — 110
- Matched example 8 Classifying walks from a list of vertices — 111
- Matched example 9 Finding the shortest path — 112
- Matched example 10 Identifying spanning trees — 113
- Matched example 11 Finding minimum spanning trees by inspection — 114
- Matched example 12 Finding minimum spanning trees using Prim's algorithm — 116

8 Variation — 117

- Matched example 1 Working with direct variation — 117
- Matched example 2 Working with inverse variation — 119
- Matched example 3 Identifying direct and inverse variation from a graph — 121
- Matched example 4 Identifying direct and inverse variation from a graph without (0, 0) — 123
- Matched example 5 Linearising data — 125
- Matched example 6 Modelling non-linear data with curves — 127
- Matched example 7 Working with non-linear data — 128

9 Measurement, scale and similarity — 129

- Matched example 1 Converting units of measurement — 129
- Matched example 2 Converting to scientific notation — 130
- Matched example 3 Converting from scientific notation — 131
- Matched example 4 Using Pythagoras' theorem to find unknown sides — 132
- Matched example 5 Using Pythagoras' theorem with shapes that contain right-angled triangles — 134
- Matched example 6 Solving problems using Pythagoras' theorem in two dimensions — 135
- Matched example 7 Solving problems using Pythagoras' theorem in three dimensions — 136
- Matched example 8 Solving two-step problems using Pythagoras' theorem in three dimensions — 137
- Matched example 9 Calculating the perimeter and area of quadrilaterals, triangles and circles — 138

Matched example 10	Calculating the perimeter and area of sectors	140
Matched example 11	Calculating the perimeter and area of composite shapes	142
Matched example 12	Calculating the area of composite shapes where a shape is removed	144
Matched example 13	Applying perimeter and area formulas	145
Matched example 14	Calculating the volume and capacity of prisms and cylinders	146
Matched example 15	Calculating the volume and capacity of pyramids, cones and spheres	147
Matched example 16	Calculating the volume of composite objects	148
Matched example 17	Using surface area formulas	150
Matched example 18	Applying surface area formulas	151
Matched example 19	Working with scale factors	152
Matched example 20	Identifying similar shapes	153
Matched example 21	Scaling areas and volumes	154

10 Appplications of trigonometry — 155

Matched example 1	Finding an unknown side of a right-angled triangle	155
Matched example 2	Solving problems involving an unknown side of a right-angled triangle	157
Matched example 3	Finding an unknown angle in a right-angled triangle	158
Matched example 4	Solving problems involving an unknown angle of a right-angled triangle	159
Matched example 5	Solving problems involving angles of elevation and depression	160
Matched example 6	Calculating three-figure bearings	161
Matched example 7	Applying three-figure bearings	163
Matched example 8	Using the sine rule for non-right-angled triangles	164
Matched example 9	Solving problems using the sine rule	166
Matched example 10	Using the cosine rule for non-right-angled triangles	168
Matched example 11	Solving problems involving non-right-angled triangles	169

Answers — 171

To the student

Nelson VICmaths is your best friend when it comes to studying General Mathematics in Year 11. It has been written to help you maximise your learning and success this year. Every explanation, every exam hack and every worked example has been written with the exams in mind.

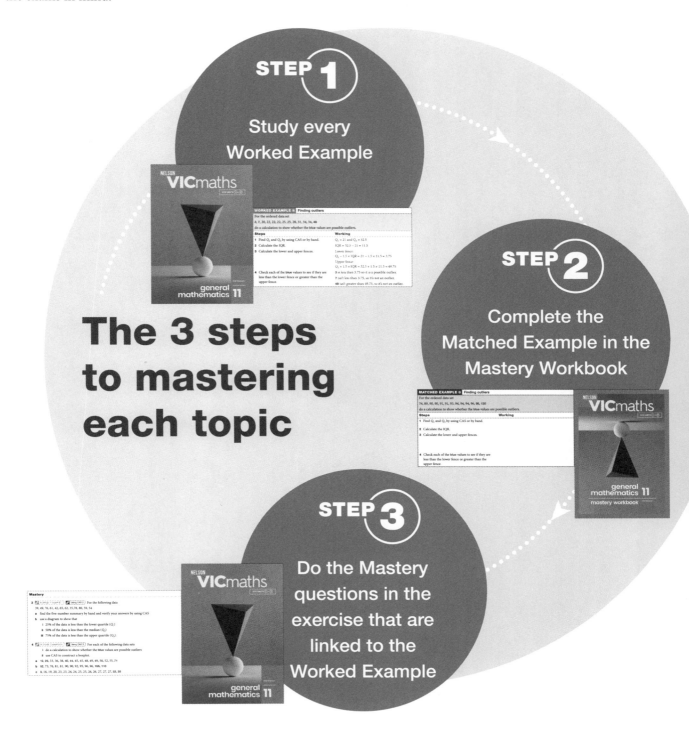

INVESTIGATING AND COMPARING DATA DISTRIBUTIONS

MATCHED EXAMPLE 1	Deciding on the type of data

State whether the following data is
  i categorical or numerical
  ii nominal, ordinal, discrete, continuous, interval or ratio.	

Steps	**Working**
a Age (in years)	

i **1** Are there numbers involved?
      2 Does it make sense to add the numbers?
  ii **1** Can you measure it with increasing levels of accuracy?
      2 Does the data scale have a fixed beginning?	

b Number of hours of sleep in a day	

i **1** Are there numbers involved?
      2 Does it make sense to add the numbers?
  ii **1** Can you measure it with increasing levels of accuracy?
      2 Does the data scale have a fixed beginning?	

c Marital status of people in Melbourne	
i **1** Are there numbers involved? ii Is there a natural order?	
d Blood group of newborns in a hospital	
i **1** Are there numbers involved? ii Is there a natural order?	
e Stress levels on a scale of 1 to 5	

i **1** Are there numbers involved?
      2 Does it make sense to add the numbers?
  ii Is there a natural order?	

▶	**f** Weight	
	i **1** Are there numbers involved?	
	2 Does it make sense to add the numbers?	
	ii **1** Can you measure it with increasing levels of accuracy?	
	2 Does the data scale have a fixed beginning?	

MATCHED EXAMPLE 2	Finding the median, mode and range

For each of the following, find the

   i median   **ii** mode   **iii** range.

Steps	Working
a Height of ten plants in a garden in metres: 2, 3, 2, 4, 1, 1, 2, 3, 6, 2	

i **1** Order the values from smallest to largest.
      2 Find the middle value (or two middle values). If there are two middle values, add them and divide by 2.
   ii Find the most commonly occurring value.
   iii range = largest value − smallest value	

b Number of cakes sold in a day by a small bakery for 11 days: 20, 21, 25, 18, 25, 24, 23, 23, 23, 22, 23	

i **1** Order the values from smallest to largest.
      2 Find the middle value (or two middle values). If there are two middle values, add them and divide by 2.
   ii Find the most commonly occurring value.
   iii range = largest value − smallest value	

MATCHED EXAMPLE 3 Constructing a frequency table

The following is raw data collected by a supermarket of the type of lolly packs that 25 people bought one day, where C = Cheekies, S = Sour Ears, F = Fads, M = Minties, J = Jaffas.

S, M, S, S, C, M, F, S, F, S, M, C, J, F, C, S, F, C, J, J, S, F, F, C, F

Set up a frequency table that includes both the number and the percentage of each type of lollies bought.

Steps

1 Set up a table with three columns and list the categories in the first column. Count the number in each category and record the frequency.

Check that the total frequency equals the total number of data values given in the question.

2 Calculate the percentage for each category using

$$\text{percentage} = \frac{\text{frequency}}{\text{total}} \times 100\%$$

Check that the total percentage is equal to 100% (or 99.9% or 100.1% if the percentages have been rounded).

Working

Lolly type	Frequency	Percentage
Cheekies		
Sour Ears		
Fads		
Minties		
Jaffas		
Total		

Answer in the above table.

MATCHED EXAMPLE 4 — Constructing a grouped frequency table

Construct a grouped frequency table for the number of runs scored by a batsman in the last 15 matches, using intervals of size 20, and find the modal interval.

32, 85, 6, 65, 52, 80, 86, 5, 45, 82, 40, 25, 26, 95, 14

Steps

1. Write the intervals in the first column.
2. Count the data values that fall into each interval.
3. Enter the frequencies in the second column.
4. Make sure you place the border values (20, 40, 60, etc.) in the correct interval.
5. Total the frequency column and check that it matches the number of data values in the list.
6. Find the interval that occurs most frequently.

Working

Runs	Frequency
Total	

MATCHED EXAMPLE 5 — Reading a bar chart

This bar chart shows the number of hours spent on Facebook over 7 days by a student.

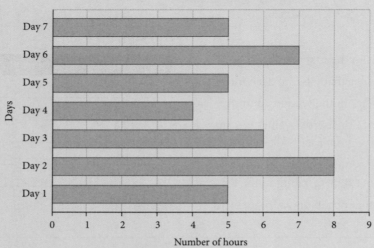

Use the graph to answer the following questions.

a On which day did the student use Facebook for the longest time?

b How many hours did the student spend on Facebook over 7 days?

c What percentage of hours did the student spend on Facebook on day 4?

Steps	Working
a Find the longest bar.	
b Add up the frequency for each day.	
c Use percentage = $\dfrac{\text{frequency}}{\text{total}} \times 100\%$	

MATCHED EXAMPLE 6 Reading and describing histograms

The following table shows the height (in cm) of a group of 110 adults that was collected for a study.

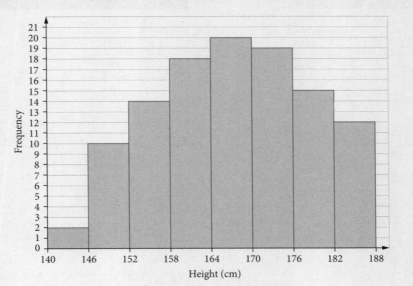

Steps	Working
a How many intervals are there?	
Count the number of intervals.	
b How many adults measured more than 1.76 m in height?	
1 Convert the m to cm. (Round to the nearest cm.)	
2 Read from the histogram.	
c What percentage of adults measured less than 1.55 m in height? Give your answer rounded to one decimal place.	
1 Convert the m to cm. (Round to the nearest cm.)	
2 Read from the histogram.	
3 Convert to a percentage using $\text{percentage} = \dfrac{\text{frequency}}{\text{total}} \times 100\%$, rounding to one decimal place.	
d Is the histogram approximately symmetric, positively skewed or negatively skewed? Does it have any possible outliers?	
The left side of the histogram closely mirrors the right side.	
e What is the modal interval?	
Which interval has the highest frequency?	

f Use the histogram to copy and complete the following grouped frequency table.

Height (kg)	Frequency
140–<146	
146–<152	
152–<158	
Total	

Height(kg)	Frequency
140–<146	
146–<152	
152–<158	
Total	

Using CAS 1:
Constructing histograms for ungrouped data
p. 22

MATCHED EXAMPLE 7 Finding the five-number summary by hand

For the following data

36, 79, 50, 49, 78, 64, 60, 52, 51, 64, 60, 59, 53, 54, 53, 55

a find the five-number summary by hand

b use a diagram to show that

   i 25% of the data is less than the lower quartile (Q_1).

   ii 50% of the data is less than the median (Q_2).

   iii 75% of the data is less than the upper quartile (Q_3).

Steps	Working
a 1 Order the data from smallest to largest.	
2 Find the minimum and maximum value.	
3 Find the median. There is an even number of data values, so we average the two middle points.	
4 Find Q_1, the median of the lower half of the data.	
5 Find Q_3, the median of the upper half of the data.	
6 List the five-number summary.	
b 1 Draw a diagram showing the three quartiles.	
2 Use the diagram to calculate the percentage of data less than each quartile.	i
	ii
	iii

MATCHED EXAMPLE 8 — Finding outliers

For the ordered data set

74, 89, 90, 90, 91, 91, 93, 94, 94, 94, 96, 98, 130

do a calculation to show whether the blue values are possible outliers.

Steps	Working
1 Find Q_1 and Q_3 by using CAS or by hand.	
2 Calculate the IQR.	
3 Calculate the lower and upper fences.	
4 Check each of the blue values to see if they are less than the lower fence or greater than the upper fence.	

Using CAS 3: Constructing boxplots p. 31

MATCHED EXAMPLE 9 Reading boxplots

The boxplot shows the distribution of 36 students' points in the qualifying round of a quiz.

Find the
a five-number summary
b percentage of students who scored more than 270
c percentage of students who scored less than 180
d percentage of students who scored between 440 and 620
e number of students who scored more than 440
f scores at the lower end that would be considered outliers
g scores at the upper end that would be considered outliers.

Steps	Working
a Read directly from the boxplot.	
Use the fact that quartiles divide data into four equal groups, so 25% of the data is in each group. 	b
	c
	d
e Find the percentage first and then multiply by the total number.	
f Use the IQR to calculate the lower fence. lower fence = $Q_1 - 1.5 \times$ IQR	
g Use the IQR to calculate the upper fence. upper fence = $Q_3 + 1.5 \times$ IQR	

MATCHED EXAMPLE 10 Using dot plots

The dot plot shows the number of books read by the members of a book club in the last month.

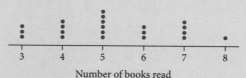

Number of books read

a Find the

  i mode
  ii range
  iii median
  iv lower quartile (Q_1)
  v upper quartile (Q_3)
  vi interquartile range (IQR).

b What could best describe the shape of the distribution: approximately symmetric, positively skewed or negatively skewed?

Steps | **Working**

a i Find the most common value.

  ii range = largest value − smallest value

  iii 1 Count the number of dots n, note whether it's odd or even and find the position of the median.

  2 If n is odd, find the data value of the middle dot.

  If n is even, find the average of the data values for the two middle dots.

  Count each column of dots from the bottom up.

  iv 1 To find the lower quartile Q_1, find the median of the lower half.

  Count the number of dots n, note whether it's odd or even and find the position of Q_1.

  2 If n is odd, find the data value of the middle dot.

  If n is even, find the average of the data values for the two middle dots.

v 1 To find the upper quartile Q_3, find the median of the upper half.

Count the number of dots n, note whether it's odd or even and find the position of Q_3.

If *n* is odd, find the data value of the middle dot.

If *n* is even, find the average of the data values for the two middle dots.

vi Use Q_3 and Q_1 to calculate the interquartile range.

b Picture the dot plot as a histogram.

MATCHED EXAMPLE 11 — Using stem plots

The stem plot shows the test marks out of 60 for a class of 33 students.

a Find the
  - **i** mode
  - **ii** range
  - **iii** median
  - **iv** lower quartile (Q_1)
  - **v** upper quartile (Q_3)
  - **vi** interquartile range (IQR)

b Is there an outlier? Justify your answer.

Stem	Leaf
2	5
3	
4	0 0 0 5 5 5 5 8 8 8
5	0 0 5 5 8 8 8 8
6	0 0 5 5 6 6
7	0 0 5 5 5
8	5 7 9
9	0 0

Key 4|0 = 40

Steps | Working

a

i Find the most common value.

ii range = largest value − smallest value

iii 1 Count the number of data values n, note whether it's odd or even and find the position of the median.

2 If n is odd, find the middle data value.
If n is even, find the average of the two middle data values.

iv 1 To find the lower quartile Q_1, find the median of the lower half.
Count the number of data values n, note whether it's odd or even and find the position of Q_1.

    2 If n is odd, find the middle data value.

    If n is even, find the average of the two middle data values.

v **1** To find the upper quartile Q_3, find the median of the upper half.

    Count the number of data values n, note whether it's odd or even and find the position of Q_3.

   2 If n is odd, find the middle data value.

    If n is even, find the average of the two middle data values.

vi Use Q_3 and Q_1 to calculate the interquartile range.

b Check any value that appears to be an outlier against the upper or lower fence.

MATCHED EXAMPLE 12 | Working with back-to-back stem plots

Two speed cameras on different roads recorded the following car speeds (in km/h).

Camera 1: 90, 87, 78, 87, 95, 70, 96, 68, 77, 65, 78, 68, 64, 88, 67, 69, 85, 79, 65, 87, 79, 68

Camera 2: 126, 137, 94, 115, 105, 136, 116, 125, 96, 124, 137, 123, 133, 118, 130, 120, 114, 108, 95, 126, 122, 114

a Display the data with a back-to-back stem plot.
b Comment on the shape of the data for Camera 2.
c Calculate the median, range and IQR for the car speeds recorded by each of the two speed cameras.
d If the speed limit on the first road is 90 km/h and on the second is 110 km/h, which road has the greater speeding problem? Provide statistical evidence for your answer.

Steps	**Working**
a Order the data and display it with a back-to-back stem plot.	Camera 1 | | Camera 2 | 6 | | 7 | | 8 | | 9 | | 10 | | 11 | | 12 | | 13 | 0|9 = 13|6 =
b To see the shape of the data in the right leaf, rotate the page 90° clockwise so that the stem forms the horizontal axis and picture it as a histogram.
c Calculate the medians, ranges and IQRs from the back-to-back stem plot.
d Compare the medians in relation to the speed limits to see if there are any noticeable differences.

MATCHED EXAMPLE 13 | Interpreting back-to-back stem plots

Mia and Chaya deliver flyers. The number of flyers delivered per hour over 14 hours is shown below.

Mia: 31, 28, 26, 30, 20, 31, 10, 20, 11, 20, 15, 16, 26, 31

Chaya: 38, 35, 28, 23, 32, 36, 23, 37, 33, 25, 34, 27, 34, 37

a Display the data with a back-to-back stem plot with split stems.
b Comment on the shape of Chaya's data.
c Calculate the median, range and IQR for the number of flyers delivered by each person.
d Who would you say is the better delivery person? Justify your answer by quoting appropriate data statistics.

Steps	Working
a Order the data and display it with a back-to-back stem plot, splitting the stems. List leaves in the range 0–4 in the first half of the split stem and leaves in the range 5–9 in the second half.	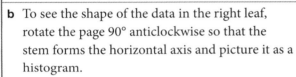 $6\|2 =$ $2\|3 =$
b To see the shape of the data in the right leaf, rotate the page 90° anticlockwise so that the stem forms the horizontal axis and picture it as a histogram.	
c Calculate the medians, ranges and IQRs from the back-to-back stem plot.	
d Use the results to decide who is the better delivery person.	

MATCHED EXAMPLE 14 Working with parallel boxplots

The parallel boxplots below represent the daily minimum temperature (°C) in October for the three towns Warwick, Forbes and Laverton.

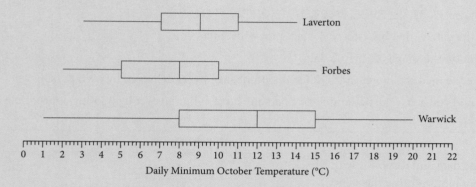

Daily Minimum October Temperature (°C)

a Which town has the lowest median average October temperature?
b Which town has the smallest range of average October temperatures?
c Which town's data is best described as negatively skewed?
d Which town had the lowest average October temperature?
e Which of the three towns has noticeably higher average October temperatures than the other two? Refer to medians as evidence in your answer.

Steps	Working
a Look for the town whose median line is nearest to the beginning of the scale.	
b Look for the smallest boxplot, including whiskers.	
c Look for the boxplot with its median to the right of the box and the left whisker being longer than the right one.	
d Look for the lowest left endpoint.	
e Compare the medians shown on the towns' boxplots.	

MATCHED EXAMPLE 15 | Calculating the mean

People of two different age groups were surveyed about the number of video games they had played in the last week. For each group

  i find how many people were surveyed
  ii calculate the mean number of video games played, correct to one decimal place.

a Group 1 video games played: 3, 5, 2, 0, 1, 6, 2, 1, 4, 8, 3, 7

b Group 2 video games played:

Number of video games (x)	Frequency
0	3
1	5
2	2
3	1
4	2
5	3

Steps | Working

a i Count the number of data values.

   ii Use the formula $\bar{x} = \dfrac{\sum x}{n}$, giving your answer correct to one decimal place.

b i Find the sum of the frequencies.

   ii 1 Add an extra column to the table and an extra row for totals. Fill in the $x \times f$ column and totals.

Number of video games (x)	Frequency	x × f
0	3	
1	5	
2	2	
3	1	
4	2	
5	3	
Total	16	

   2 Use the formula $\bar{x} = \dfrac{\sum xf}{\sum f}$, giving your answer correct to one decimal place.

Using CAS 5:
Finding the mean and standard deviation for ungrouped data
p. xx

Using CAS 6:
Finding the mean and standard deviation for grouped data p. xx

MATCHED EXAMPLE 16 | Working with the mean and standard deviation from a display

For each of the following, find
  i the mean and standard deviation, correct to two decimal places.
  ii the number of data values that are one standard deviation from the mean.

Steps	Working
a 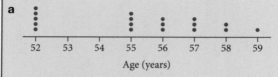	
i Use CAS by entering the data values from the graph and selecting the mean and standard deviation. Round to two decimal places.	
ii 1 Find $\bar{x} - s$ and $\bar{x} + s$.	
2 Count the number of data values between $\bar{x} - \sigma s$ and $\bar{x} + s$.	
b

4	0 1 2 3 4
4	5 5 5 7
5	0 1 2
5	5 5 6

Key 4\|5 = 45 years | |
i Use CAS by entering the data values from the graph and selecting the mean and standard deviation. Round to two decimal places.	
ii 1 Find $\bar{x} - s$ and $\bar{x} + s$.	
2 Count the number of data values between $\bar{x} - s$ and $\bar{x} + s$.	

MATCHED EXAMPLE 17 | Working with the mean-standard deviation scale

A study found that the number of chocolate chips in a box of ice cream has a normal distribution with a mean of 24 and a standard deviation of 6. Find the percentage of ice cream boxes that have

a between 24 and 36   **b** more than 36   **c** between 12 and 18   **d** less than 6

chocolate chips.

Steps	Working
a 1 Write a mean-standard deviation scale that includes the mean and standard deviation values given.	
2 Add the required percentages from the mean-standard deviation scale.	
b Add the required percentages from the mean-standard deviation scale.	
c Read from the mean-standard deviation scale.	
d Read from the mean-standard deviation scale.	

CHAPTER 2
ARITHMETIC SEQUENCES AND FINANCIAL RECURRENCE RELATIONS

MATCHED EXAMPLE 1 Finding sequences from rules

For each of the following sequence rules
  i find the first five values
  ii describe in words how each new value is generated.

a $u_n = 3n - 1$, where $n = 0, 1, 2 \ldots$
b $u_n = 2 + 4n$, where $n = 0, 1, 2 \ldots$
c $u_n = 3^{n-1}$, where $n = 1, 2, 3 \ldots$
d $V_n = (-2)^n$, where $n = 0, 1, 2 \ldots$

Steps	Working
a i Substitute $n = 0, 1, 2, 3, 4$ into the rule.	
ii Find the pattern from one value to the next.	
b i Substitute $n = 0, 1, 2, 3, 4$ into the rule.	
ii Find the pattern from one value to the next.	
c i Substitute $n = 0, 1, 2, 3, 4$ into the rule.	
ii Find the pattern from one value to the next.	

d **i** Substitute $n = 0, 1, 2, 3, 4$ into the rule.

  ii Find the pattern from one value to the next.

MATCHED EXAMPLE 2 Identifying sequence types

Identify whether each of the following are increasing, decreasing, constant, oscillating or limiting value sequences, giving more than one option where relevant.

a

n	0	1	2	3	4	...
u_n	−1	1	−1	1	−1	...

b

n	0	1	2	3	4	...
u_n	2	4	8	16	32	...

c

n	1	2	3	4	5	6	7	8	...
u_n	1	0.5	0.33	0.25	0.2	0.17	0.14	0.13	...

d
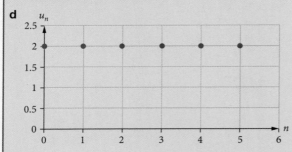

e
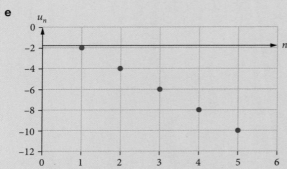

Steps	**Working**
a Are the values always increasing or decreasing?	
b Do the values stay the same?	
c Do the values alternate between a positive number and a negative number?	
d Do the values tend towards a particular value without ever reaching it?	
e	

MATCHED EXAMPLE 3 Graphing arithmetic sequences

For each of the following sequences

a 7, 10, 13, 16 …

b 4, 2, 0, −2 …

  i explain why it's an arithmetic sequence

  ii find u_2 and u_5

  iii make a table of values showing all the values of u_n up to $n = 6$

  iv sketch a graph of the table of values in part **iii**

  v find the slope of the straight line created by joining the points and comment on the result.

Steps	Working
a **i** Is a fixed value being added or subtracted each time?	
ii Extend the sequence by continuing the rule if necessary.	
iii List n in the first row of the table and u_n in the second. Find u_n for $n = 0$ to 6.	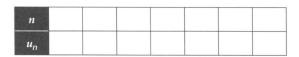
iv Sketch the table of values with n on the horizontal axis and u_n on the vertical axis.	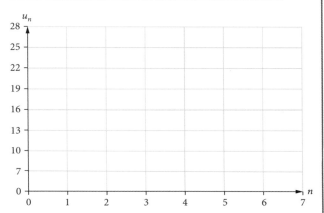
v Use any two points to calculate the slope and compare it to the sequence. Often the first two points are the easiest to use.	
b **i** Is a fixed value being added or subtracted each time?	
ii Extend the sequence by continuing the rule if necessary.	
iii List n in the first row of the table and u_n in the second. Find u_n for $n = 0$ to 6.	

iv Sketch the table of values with *n* on the horizontal axis and u_n on the vertical axis.

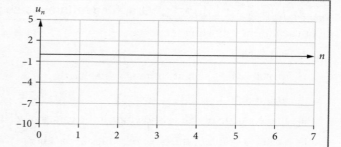

v Use any two points to calculate the slope and compare it to the sequence. Often the first two points are the easiest to use.

MATCHED EXAMPLE 4 Finding arithmetic sequences from recurrence relations

For the recurrence relation $u_0 = 2$, $u_{n+1} = u_n + 3$, find

a a and d

b the first four values of the arithmetic sequence generated by the recurrence relation, showing all the calculation steps.

Steps	Working
a Use $u_0 = a$, $u_{n+1} = u_n + d$ where a is the first value d is the common difference between values.	
b 1 The first part of the recurrence relation gives us u_0.	
2 The second part of the recurrence relation tells us what to add to generate the next value.	
3 Repeat for the next two values.	
4 Write down the first four values of the sequence.	

MATCHED EXAMPLE 5 — Finding the nth value of an arithmetic sequence

For each of the following sequences

i explain how we know that it is an arithmetic sequence

ii find the nth value rule

iii use the rule to find u_{20} and u_{100}.

Steps	Working
a 1, 5, 9, 13, …	
i Are we adding or subtracting the same value to generate each new value?	
ii 1 Find the first value a and the common difference d.	
2 Substitute the values for a and d into the nth value rule for arithmetic sequence $u_n = a + nd$ and simplify.	
iii Substitute the values for n into the nth value rule of the sequence $u_n = a + nd$.	
b $u_0 = 60$, $u_{n+1} = u_n - 3$	
i Are we adding or subtracting the same value to generate each new value?	
ii 1 Find the first value a and the common difference d.	
2 Substitute the values for a and d into the nth value rule for arithmetic sequence $u_n = a + nd$ and simplify.	
iii Substitute the values for n into the nth value rule of the sequence $u_n = a + nd$.	

MATCHED EXAMPLE 6 — Using simple interest recurrence relations

Mia invests $2000 in an account earning 6% per annum simple interest.

Steps | **Working**

a What is the fixed amount of interest paid for each year?

Use $d = \dfrac{r}{100} \times u_0$ to find the fixed amount of interest each year. Round the answer if required.

b Copy and complete the following table to find
  i Mia's bank account balance after three years
  ii the first year when her balance is greater than $2400
  iii the total amount of interest earned after six years.

n	Account balance after n years ($)
0	2000
1	2000 + =
2	+ =
3	+ =
4	+ =
5	+ =
6	+ =

Complete the table by using CAS recursive computation to find the bank account balance after six years.

n	Account balance after n years ($)
0	2000
1	2000 + =
2	+ =
3	+ =
4	+ =
5	+ =
6	+ =

TI-Nspire   **ClassPad**

  i Read the answer from the table.

    ii Read the answer from the table.

    iii Total amount of interest earned after n years $= u_n - u_0$.

c Write a recurrence relation, u_n, for the account balance for n years.

Identify the starting value. Each value is calculated by adding d to the previous value.

d Sketch the graph of the recurrence relation up to $n = 6$ and describe the line made by the points.

The horizontal axis is n (years) and the vertical axis is u_n ($).

Plot the values from the table.

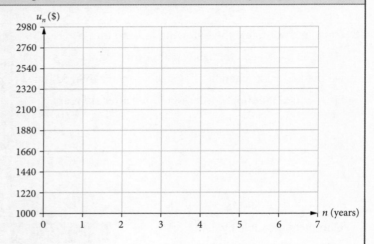

MATCHED EXAMPLE 7 — Using recursive computation for simple interest

Find the balance after six years for each of the following using CAS recursive computation.

a Maya invests $5000 in a bank account earning 4% per annum simple interest.

b Mathew takes out a loan of $5000 from a bank at 4% per annum simple interest.

Steps	Working
a 1 Use $d = \dfrac{r}{100} \times u_0$ to find the fixed amount of interest for each year. Add d for an investment and subtract d for a loan.	
2 Use CAS recursive computation to find the balance after six years. **TI-Nspire** **ClassPad**	
3 Write the answer.	
b 1 Use $d = \dfrac{r}{100} \times u_0$ to find the fixed amount of interest for each year. Add d for an investment and subtract d for a loan.	
2 Use CAS recursive computation to find the balance after six years. **TI-Nspire** **ClassPad**	
3 Write the answer.	

MATCHED EXAMPLE 8 Using the simple interest general rule

Marcus invests $3000 in an account earning 6% p.a. simple interest.

Steps	Working
a Find the fixed amount of interest paid each year.	
Find the value of d, using $d = \dfrac{r}{100} \times u_0$, rounding the answer if required.	
b Write a rule that will calculate the balance of the account after n years.	
Decide if this is an investment or a loan and choose either $u_n = u_0 + nd$ or $u_n = u_0 - nd$.	
c Use the rule to find the balance of the account after ten years.	
Substitute the value of n into the rule.	

MATCHED EXAMPLE 9 Using flat rate depreciation recurrence relations

A car is purchased for $20 000. Its value depreciates at a flat rate of 20% each year.

Steps	Working
a What is the fixed amount of depreciation each year?	
Use $d = \dfrac{r}{100} \times u_0$ to find the fixed amount of depreciation each year.	

b Copy and complete the following table to find

  i the value of the car after three years

  ii when the value of the car first falls below $10 000

  iii when the car depreciates to zero.

n	Value after n years ($)
0	20 000
1	20 000 − =
2	− =
3	− =
4	− =
5	− =

1 Complete the table by using CAS recursive computation to find the value of the asset after five years.

n	Value after n years ($)
0	20 000
1	20 000 − =
2	− =
3	− =
4	− =
5	− =

2 **i** Read the answer from the table.

  ii Read the answer from the table.

  iii Read the answer from the table.

c Write a recurrence relation for the value of the car.

Identify the initial value of the asset, u_0.
Each value is calculated by subtracting d from the previous value.

d Sketch the graph of the recurrence relation up to $n = 5$.

The horizontal axis is n (years) and the vertical axis is u_n ($).

Plot the values from the table.

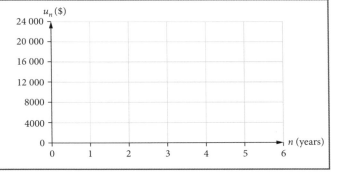

MATCHED EXAMPLE 10 Finding the rate for flat rate depreciation

A restaurant owns a commercial refrigerator that is depreciated in value using flat rate depreciation. The value of the refrigerator, in dollars, after n years, u_n, can be modelled by the recurrence relation

$$u_0 = 10\,500, \qquad u_{n+1} = u_n - 1050$$

a By what amount, in dollars, does the value of the refrigerator decrease each year?

b Showing recursive calculations, determine the value of the refrigerator, in dollars, after three years.

c What annual flat rate percentage of depreciation is used by the restaurant?

Steps	Working
a Use the recurrence relation $u_{n+1} = u_n - d$ to identify d.	
b Show the stepped calculations for u_1, u_2 and u_3 from the recurrence relation.	
c 1 Identify what we know and what we need to find. 2 Substitute the values into $r = \dfrac{d}{u_0} \times 100\%$ and evaluate. 3 Write the answer.	

MATCHED EXAMPLE 11 — Using the flat rate depreciation general rule

A construction company buys a bulldozer for $383 000 and depreciates it at a flat rate of 16% per year.

Steps	Working
a Find the fixed amount of depreciation each year.	
Use $d = \dfrac{r}{100} \times u_0$ to find the fixed amount of depreciation each year.	
b Write a rule that will calculate the value of the bulldozer after n years.	
Substitute the values of d and u_n into the flat rate depreciation general rule $u_n = u_0 - nd$.	
c Use the rule to find the value of the bulldozer after four years.	
Substitute the value of n into the rule.	
d Use the rule to find how many years it would take for the bulldozer to depreciate to zero.	
1 Substitute the known values into $u_n = u_0 - nd$. Let $u_n = 0$.	
2 Solve for n, using CAS if necessary.	
3 Write the answer. When solving for n always round *up* to the nearest year.	

MATCHED EXAMPLE 12 Using unit cost depreciation recurrence relations

A garment factory buys a sewing machine for $5000 that depreciates by $50 every time it produces a garment.

Steps	Working

a Explain why this is unit cost depreciation not flat rate depreciation.

1 Refer to 'use' in the answer.

b Copy and complete the following table to find

i the value of the sewing machine after it produces four garments

ii how many garments will it take for the value of the sewing machine to first fall below $4800.

n	Value after n years ($)
0	5000
1	5000 − =
2	− =
3	− =
4	− =
5	− =

1 Complete the table by using CAS recursive computation to find the value after four garments.

n	Value after n years ($)
0	5000
1	5000 − =
2	− =
3	− =
4	− =
5	− =

2 i Read the answer from the table.

ii Read the answer from the table.

c Write a recurrence relation for the value of the sewing machine.

Identify the initial value of the asset and the cost per unit of use.

d Sketch the graph of the recurrence relation up to $n = 5$.

The horizontal axis is n (units of use) and the vertical axis is u_n ($).

Plot the values from the table.

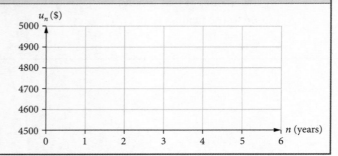

MATCHED EXAMPLE 13 — Using the unit cost depreciation general rule

A photocopier was purchased for $2000. The photocopier's value depreciates at a rate of 50 cents per copy.

Steps	Working
a Write a rule that will calculate the value of the photocopier after n copies.	
Substitute the values of d and u_0 into the unit cost depreciation general rule $u_n = u_0 - nd$.	
b Use the rule to find the value of the photocopier after it has produced 100 copies.	
Substitute the value of n into the rule.	

CHAPTER 3
GEOMETRIC SEQUENCES AND FINANCIAL MATHEMATICS

p. 118

MATCHED EXAMPLE 1 | Graphing geometric sequences

For each of the following sequences
a 55, 110, 220, 440 …
b 54, 18, 6, 2 …

  i explain why it's a geometric sequence
  ii show three calculations for finding the common ratio, R.
  iii find u_2 and u_4
  iv make a table of values, showing all the values of u_n for $n = 0$ to 5
  v sketch a graph of the table of values in part **iv**.

Steps	Working
a i Is a fixed value being multiplied each time?	
ii Use $R = \dfrac{\text{any value}}{\text{previous value}}$	
iii Extend the sequence by continuing the rule if necessary.	
iv List n in the first row of the table and u_n in the second. Find u_n for $n = 0$ to 5.	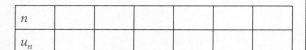
v Sketch the table of values with n on the horizontal axis and u_n on the vertical axis.	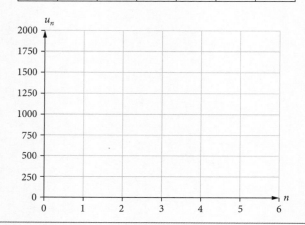

b i Is a fixed value being multiplied each time?

ii Use $R = \dfrac{\text{any value}}{\text{previous value}}$

iii Extend the sequence by continuing the rule if necessary.

iv List n in the first row of the table and u_n in the second. Find u_n for $n = 0$ to 5.

n						
u_n						

v Sketch the table of values with n on the horizontal axis and u_n on the vertical axis.

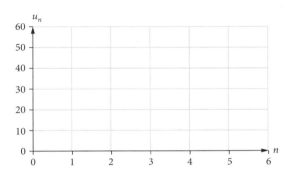

p. 119

MATCHED EXAMPLE 2 — Finding geometric sequences from recurrence relations

For the recurrence relation $u_0 = 3$, $u_{n+1} = -4u_n$, find

a a and R

b the first four values of the geometric sequence generated by the recurrence relation, showing all the calculation steps.

Steps	Working
a Use $u_0 = a$, $u_{n+1} = Ru_n$ where a is the first value R is the common ratio between values	
b 1 The first part of the recurrence relation gives you u_0. **2** The second part of the recurrence relation tells you what to add to generate the next value. **3** Repeat for the next two values. **4** Write down the first four values of the sequence.	

Using CAS 1: Generating geometric sequences through recursive computation p. 120

MATCHED EXAMPLE 3 Finding the nth value of a geometric sequence

For each of the following sequences
- i explain how we know that it is a geometric sequence
- ii find the nth value rule
- iii use the rule to find u_5 and u_8, rounding to two decimal places if necessary.

Steps	Working
a 16, 64, 256, 1024, …	
i Are we multiplying the same value to generate each new value?	
ii 1 Find the first value a and the common ratio R.	
2 Substitute the values for a and R into the nth value rule for geometric sequences $u_n = aR_n$.	
iii Substitute the values for n into the nth value rule of the sequence, rounding to two decimal places if necessary.	
b $u_0 = -500$, $u_n + 1 = 0.5u_n$	
i Are we multiplying the same value to generate each new value?	
ii 1 Find the first value a and the common ratio R.	
2 Substitute the values for a and R into the nth value rule for geometric sequences $u_n = aR_n$.	
iii Substitute the values for n into the nth value rule of the sequence, rounding to two decimal places if necessary.	

MATCHED EXAMPLE 4 | Comparing compound and simple interest

Ariana is investing $8000 for four years and wants to compare an investment at 15% p.a (per year) compounding yearly to 15% p.a. simple interest.

a Copy and complete the following table for $n = 3$ and $n = 4$.

	Compound		Simple	
n	Interest ($)	Value of investment ($)	Interest ($)	Value of investment ($)
0	–	8000	–	8000
1	$\frac{15}{100} \times 8000 = 1200$	$8000 + 1200 = 9200$	$\frac{15}{100} \times 8000 = 1200$	$8000 + 1200 = 9200$
2	$\frac{15}{100} \times 9200 = 1380$	$9200 + 1380 = 10\,580$	$\frac{15}{100} \times 8000 = 1200$	$9200 + 1200 = 10\,400$
3				
4				

b What is the value of the compound interest investment after four years?

c After four years, how much more is the value of the compound interest investment compared to the simple interest investment?

Steps	Working

a

	Compound		Simple	
n	Interest ($)	Value of investment ($)	Interest ($)	Value of investment ($)
0	–	8000	–	8000
1	$\frac{15}{100} \times 8000 = 1200$	$8000 + 1200 = 9200$	$\frac{15}{100} \times 8000 = 1200$	$8000 + 1200 = 9200$
2	$\frac{15}{100} \times 9200 = 1380$	$9200 + 1380 = 10\,580$	$\frac{15}{100} \times 8000 = 1200$	$9200 + 1200 = 10\,400$
3				
4				

b Read from the table.

c Compare the last entries in the two interest ($) columns in the table.

MATCHED EXAMPLE 5 | Working with compounding periods

For each of the following investments, find
　i　the number of compounding periods per year
　ii　the number of compounding periods over ten years
　iii　the percentage interest rate per compounding period, written as a fraction
　iv　the amount of interest earned in the first compounding period to the nearest cent.

a Kim invests $15 000 at 6% compound interest per annum compounding monthly.
b Katrina invests $156 000 at 10% compound interest per annum compounding weekly.

Steps	Working
a　i　How many compounding periods are there per year?	
ii　Multiply the number of compounding periods per year by the number of years.	
iii　Divide the percentage interest rate per year by the number of compounding periods per year.	
iv　Convert the compounding period interest rate to a decimal and multiply by the principal. Round to the nearest cent.	
b　i　How many compounding periods are there per year?	
ii　Multiply the number of compounding periods per year by the number of years.	
iii　Divide the *percentage interest rate per year* by the number of compounding periods per year.	
iv　Convert the compounding period interest rate to a decimal and multiply by the principal. Round to the nearest cent.	

MATCHED EXAMPLE 6 — Finding compound interest recurrence relations

Write a recurrence relation for the account balance, after n compounding periods, for each of the following in simplest form.

a Cathy deposited $25 000 into a savings account earning compound interest at the rate of 4.4% per annum, compounding quarterly.

b Cora deposited $30 000 into a savings account earning compound interest at the rate of 5.2% per annum, compounding weekly.

c Caitlin deposited $35 000 into a savings account earning compound interest at the rate of 7.3% per annum, compounding daily.

Steps	Working
a 1 Find the number of compounding periods per year. 2 Identify u_n, u_0 and r. 3 Substitute the values into $u_0 =$ principal, $u_{n+1} = \left(1 + \dfrac{r}{100}\right)u_n$ and simplify.	
b 1 Find the number of compounding periods per year. 2 Identify u_n, u_0 and r. 3 Substitute the values into $u_0 =$ principal, $u_{n+1} = \left(1 + \dfrac{r}{100}\right)u_n$ and simplify.	
c 1 Find the number of compounding periods per year. 2 Identify u_n, u_0 and r. 3 Substitute the values into $u_0 =$ principal, $u_{n+1} = \left(1 + \dfrac{r}{100}\right)u_n$ and simplify.	

MATCHED EXAMPLE 7 | Using the compound interest rule

Maria invests $45 000 in an account where she earns interest of 4% p.a. compounded quarterly.

a Find r, the percentage interest rate per compounding period.
b Write a rule that will calculate the value of the investment after n compounding periods.
c Use the rule to find the value of the investment after 20 compounding periods to the nearest cent.
d Calculate the total amount of interest paid after 20 compounding periods to the nearest cent.

Steps	Working
a Divide the yearly interest rate by the number of compounding periods per year.	
b Substitute the values of u_0 and r into the compound interest general rule $u_n = \left(1 + \dfrac{r}{100}\right)^n \times u_0$ and simplify.	
c 1 Substitute the value of n, the number of compounding periods, into the rule. 2 Write the answer, rounding to the nearest cent.	
d Total amount of interest after n compounding periods $= u_n - u_0$. Write the answer, rounding to the nearest cent.	

MATCHED EXAMPLE 8 — Finding reducing balance depreciation recurrence relations

A school administrator purchased a school bus for $70 000. It is depreciated using reducing balance depreciation at a rate of 20% per annum. Give all answers to the nearest dollar.

a Copy and complete the table to find the value of the school bus after five years.

n	Depreciation after n years ($)	Value after n years ($)
0	–	70 000
1	$\frac{20}{100} \times 70\,000 = 14\,000$	$70\,000 - 14\,000 = 56\,000$
2	$\frac{20}{100} \times 56\,000 = 11\,200$	$56\,000 - 11\,200 = 44\,800$
3		
4		
5		

b How much has the school bus depreciated in the third year?

c Write down a recurrence relation that gives the value of the school bus after n years.

Steps	Working
a 1 Calculate the percentage of successive values and subtract from the previous value. Use CAS recursive computation where possible. Give all values to the nearest dollar, but don't round until after all the calculations have been done. [Note: Answers can vary slightly depending on when values are rounded.]	*(table with columns: n, Depreciation after n years ($), Value after n years ($); rows 0–5 with rows 0,1,2 filled as above and rows 3,4,5 blank)*
2 Read the answer from the table.	
b Read the answer from the table.	
c 1 Identify u_n, u_0 and r.	
2 Substitute the values into u_0 = initial value of the asset, $u_{n+1} = \left(1 - \dfrac{r}{100}\right)u_n$ and simplify.	

MATCHED EXAMPLE 9 | Using reducing balance depreciation recurrence relations

A truck is depreciated using the reducing balance method of depreciation. The value of the truck u_n, in dollars, after n years, can be modelled by the recurrence relation $u_0 = 85\,000$, $u_{n+1} = 0.8u_n$.

a At what annual percentage rate is the value of the truck depreciated each year?

b When does the truck first depreciate to under $45\,000?

Steps	Working
a 1 Use u_0 = initial value of the asset, $$u_{n+1} = \left(1 - \frac{r}{100}\right) \times u_n$$ 2 Solve for r, using CAS if necessary. **TI-Nspire** **ClassPad** Write the answer.	
b Use CAS recursive computation. **TI-Nspire** **ClassPad** Find when the value is first below $45\,000.	

MATCHED EXAMPLE 10 — Using the reducing balance depreciation rule

A sports bar bought a TV for $8000 and is depreciating it by 10% of its value each year.

Steps	Working
a Explain why this involves reducing balance depreciation and not flat rate depreciation. Does the depreciation involve a changing amount or a fixed amount each year?	
b Write a rule that will calculate the value of the TV after n years. Substitute the values of u_0 and r into the reducing balance depreciation general rule $u_n = \left(1 - \dfrac{r}{100}\right) \times u_0$ and simplify.	
c Use the rule to find the value of the TV after six years to the nearest dollar. Substitute the value of n into the rule and solve.	

MATCHED EXAMPLE 11 — Calculating mark-ups and discounts

The original price of a pair of shoes was $95.

a What is the change in price if the pair of shoes is
  i marked up by 15%
  ii discounted by 10%.

b Find the new price if the pair of shoes is
  i marked up by 6%
  ii discounted by 12%.

Steps	Working
a i change in price = original price $\times \dfrac{r}{100}$ **ii** change in price = $-\left(\text{original price} \times \dfrac{r}{100}\right)$	
b i new price = original price $+ \left(\text{original price} \times \dfrac{r}{100}\right)$ **ii** new price = original price $- \left(\text{original price} \times \dfrac{r}{100}\right)$	

MATCHED EXAMPLE 12 | Finding the percentage change *r*

The original price of a pair of shoes was $95.
a Find the percentage mark-up if it has been increased to $114.
b Find the percentage discount if it has been discounted to $80.75.

Steps	Working
a $r = \dfrac{\text{new price} - \text{original price}}{\text{original price}} \times 100\%$	
b $r = \dfrac{\text{original price} - \text{new price}}{\text{original price}} \times 100\%$	

MATCHED EXAMPLE 13 Finding the original price

Find the original price of a pair of shoes to the nearest cent if

a the new price is $100 and the mark-up is 10%

b the new price is $65 and the discount is 15%.

Steps	Working
a 1 original price = new price × $\dfrac{100}{(100+r)}$	
2 Give the answer to the nearest cent.	
b 1 original price = new price × $\dfrac{100}{(100-r)}$	
2 Give the answer to the nearest cent.	

MATCHED EXAMPLE 14 — Working with GST

Round the following answers to the nearest dollar.

a A truck is advertised at a price of $75 000, which includes GST.
   i What is the price excluding GST?
   ii How much of the advertised price is GST?

b An oven has a GST-free price of $140.
   i What is the price with GST?
   ii How much GST is payable?

Steps	Working
a i Use price without GST $= \dfrac{\text{price with GST}}{1.1}$, rounding to the nearest dollar.	
ii Use GST amount $= \dfrac{\text{price with GST}}{11}$, rounding to the nearest dollar.	
b i Use price with GST = price without GST $\times$ 1.1, rounding to the nearest dollar.	
ii Use GST amount = price without GST $\times$ 0.1, rounding to the nearest dollar.	

MATCHED EXAMPLE 15 | Using the unitary method

Answer the following questions using the unitary method.

a If 14 burgers cost $173.60, find the cost of
  i one burger   **ii** 50 burgers.

b If 8 packets of flour have a total mass of 4200 grams, find the mass of
  i one packet   **ii** 30 packets.

Steps	Working
a **i** 1 Write the known equality.	
2 Identify the quantity being asked for.	
3 Find the cost of one burger: the unit amount.	
4 Write the answer in words.	
ii 1 Multiply the unit amount by the required number.	
2 Write the answer in words.	
b **i** 1 Write the known equality.	
2 Identify the quantity being asked for.	
3 Find the mass of one packet: the unit amount.	
4 Write the answer in words.	
ii 1 Multiply the unit amount by the required number.	
2 Write the answer in words.	

MATCHED EXAMPLE 16 — Using the unitary method for percentages

An apartment block has 156 women residents and women make 52% of the residents.

a What is the total number of residents in the apartment block?

b If 10% of the apartment block's residents are children, how many children are there?

Steps	Working
a 1 Write the known equality.	
2 Identify the quantity being asked for.	
3 Divide this amount by the number of units to find the unit amount (i.e. 1% of the population).	
4 Multiply the unit amount by 100 to find the whole amount (i.e. 100% of the population).	
5 Write the answer in words.	
b 1 Multiply the unit amount by the required number.	
2 Write the answer in words.	

MATCHED EXAMPLE 17 — Using the unitary method to make comparisons

Sam is buying coffee beans. His brand comes in three different sizes: 250 g for $5.35, 500 g for $9.70 and $5.35 and 900 g for $17.55.

a What is the unit cost of each of the sizes? Give your answer correct to four decimal places.

b Which size is the best value?

Steps	Working
a 1 Write the known equality for each of the sizes. If necessary, convert units of measurement so that all the units are the same. 2 Identify the quantity being asked for. 3 Divide this quantity to find the unit amounts for each of the sizes (i.e. the cost of one gram). Give your answer correct to four decimal places.	
b 1 Compare to find the cheapest and write the answer in words.	

MATCHED EXAMPLE 18 | Finding single item inflation rates

What is the single item inflation rate *r* that measures the change in price of the following items over the time periods given? Give your answer rounded to one decimal place.

a A litre of milk cost 3 cents in 1901 and $1.63 in 2022.

b A dozen eggs cost 60 cents in 1972 and $5.18 in 2022.

Steps	Working
a 1 Use $r = \dfrac{\text{new price} - \text{original price}}{\text{original price}} \times 100\%$ 2 Give your answer to rounded one decimal point.	
b 1 Use $r = \dfrac{\text{new price} - \text{original price}}{\text{original price}} \times 100\%$ 2 Give your answer to rounded one decimal point.	

MATCHED EXAMPLE 19 | Working with inflation rates

A country has an official inflation rate of 2.8% in 2023, 3.4% in 2024 and 3.6% in 2025. If the price of 1 kg of sugar at the start of 2023 is $1.15, and its price rises match the official inflation rate exactly, show the calculation that will give the price of 1 kg of sugar to the nearest cent

a at the end of 2023

b at the end of 2024

c at the end of 2025

d What is the single item inflation rate of 1 kg of sugar from the start of 2023 to the end of 2025, rounded to one decimal point?

e Show that the single item inflation rate is not the same as adding together the inflation rates in each of the three years, and explain why this is so.

p. 146

Steps	Working
a 1 Use the recurrence relation $u_0 =$ original price, $u_{n+1} = \left(1 + \dfrac{r}{100}\right) u_n$.	
2 Write the answer, rounding to the nearest cent.	
b 1 Use the recurrence relation $u_0 =$ original price, $u_{n+1} = \left(1 + \dfrac{r}{100}\right) u_n$. Use the unrounded previous answer.	
2 Write the answer, rounding to the nearest cent.	
c 1 Use the recurrence relation $u_0 =$ original price, $u_{n+1} = \left(1 + \dfrac{r}{100}\right) u_n$. Use the unrounded previous answer.	
2 Write the answer, rounding to the nearest cent.	
d 1 Use $r = \dfrac{\text{new price} - \text{original price}}{\text{original price}} \times 100\%$	
2 Give your answer to rounded one decimal point.	
e 1 Sum the inflation rates in each of the three years and compare this with the single item inflation rate.	
2 Refer to the compounding nature of inflation.	

MATCHED EXAMPLE 20 | Calculating the purchasing power of money

Amala has kept $210 in a box for ten years. Over that time, inflation has averaged 3.6% each year. Find the purchasing power of the $210 after ten years to the nearest cent?

Steps	Working
a 1 Since the inflation rate is given as an average over a number of years, use the reducing balance depreciation rule $$u_n = \left(1 - \frac{r}{100}\right)^n \times u_0$$	
2 Write you answer, rounding to the nearest cent.	

MATCHED EXAMPLE 21 | Understanding personal loans

Hugo has taken out a two-year $5000 personal loan at 5% per annum, compounding quarterly, with annual payments of $444.25.

a Find the quarterly interest rate.

Show the following calculations to the nearest cent.

b What is the balance of the loan

    i after one quarter?   **ii** after two quarters?   **iii** after three quarters?

c Find the total amount of money Hugo will have paid on this loan at the end of the three years.

d What is the total interest Hugo will pay on this loan?

Steps	Working
a Divide the yearly interest rate by the number of compounding periods per year.	
b new balance = old balance + $\left(\dfrac{r}{100} \times \text{old balance}\right)$ – regular payment, where r is the percentage interest rate per compounding period.	**i** **ii** **iii**
c total amount paid = number of payments × regular payment amount	
d total interest = total amount paid – loan amount	

MATCHED EXAMPLE 22 — Comparing purchasing options

Anika wants to buy a video console. She doesn't have the $1100, so she is looking at her options. Calculate the cost of each of the following over one year to the nearest dollar.

a Anika decides to wait three months until she has enough cash to pay for the video console, and it ends up being on sale at a 15% discount.

b Anika gets a debit card with a $3 monthly bank fee, and the seller has a 2% merchant surcharge.

c Anika uses a buy now pay later service (BNPL) of six fortnightly payments but ends up paying an $80 late fee.

d Anika gets a credit card and ends up paying 18% per annum interest for 60 days, compounding monthly, because she couldn't make the payment.

e Anika takes out a personal loan at 15% per annum compound interest, compounding monthly, which involves making six monthly payments of $191.44.

f Which option has the lowest cost?

Steps	Working
a 1 For an item discounted by $r\%$. New price = original price $- \left(\text{original price} \times \dfrac{r}{100}\right)$ **2** State the answer to the nearest dollar.	
b 1 For an item marked up by $r\%$. New price = Original price $+ \left(\text{Original price} \times \dfrac{r}{100}\right)$ **2** Calculate the total bank fee for the year. **3** State the answer to the nearest dollar.	
c Add the late fee to the price.	
d 1 Divide the yearly interest rate by the number of compounding periods per year. **2** Use the interest general rule for compound interest $u_n = \left(1 + \dfrac{r}{100}\right)^n \times u_0$ **3** State the answer to the nearest dollar.	
e Multiply the payment by the number of months.	
f Which cost is the lowest?	

CHAPTER 4
LINEAR FUNCTIONS, GRAPHS, EQUATIONS AND MODELS

MATCHED EXAMPLE 1 Drawing linear graphs from tables of values

For the linear function $y = 2x - 7$
a find the y values for each of $x = -2, -1, 0, 1, 2$
b use the y values to construct a table of values for the linear function
c use the table of values to draw the linear function by hand.

Steps	Working
a Use the linear function to calculate the y value for each x value.	
b Set up a table of values for the linear function, to show the x values and their corresponding y values.	
c Draw a Cartesian plane and plot the points from the table. Draw a line through the points, extending it and placing arrows on both ends of the line. Label the graph with its equation.	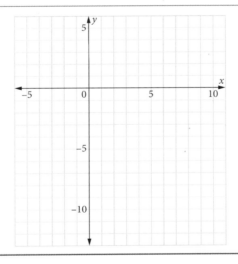

Using CAS 1:
Drawing linear graphs and generating tables of values
p. 164

MATCHED EXAMPLE 2 — Determining if a point lies on a line

Determine if the point (3, 14) lies on the following lines, showing a calculation to justify your answer.

a $y = -5x + 3$ b $y = 5x - 1$

Steps	Working
a 1 Substitute the coordinates into the equation of the line, evaluate and state whether the equation is true or false.	
2 Write the answer.	
b 1 Substitute the coordinates into the equation of the line, evaluate and state whether the equation is true or false.	
2 Write the answer.	

MATCHED EXAMPLE 3 — Finding the slope using $\frac{\text{rise}}{\text{run}}$

For each of the lines shown below
  i state whether the slope is positive, negative, zero or not defined
  ii if it is positive or negative, calculate the slope of the line using $\frac{\text{rise}}{\text{run}}$ for the two points shown.

a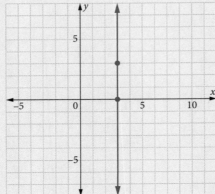

b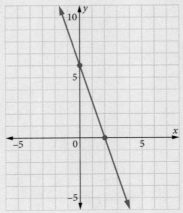

c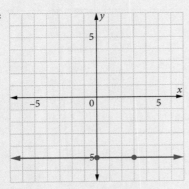

d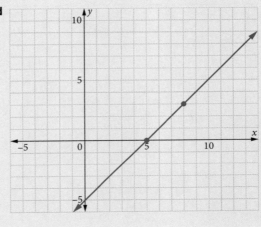

Steps	Working
a i The line is vertical.	
b i Is the line sloping up or down from left to right?	
ii 1 Draw in a right-angled triangle using the two points and find the rise and run between them.	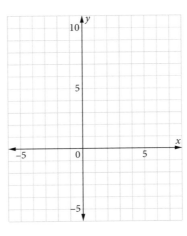
2 Use slope $=\frac{\text{rise}}{\text{run}}$ and simplify.	

c i The line is horizontal.

d i Is the line sloping up or down from left to right?

   ii 1 Draw in a right-angled triangle using the two points and find the rise and run between them.

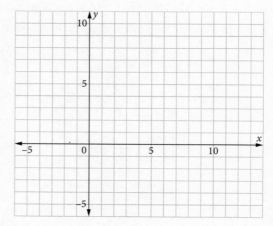

     2 Use slope $= \dfrac{\text{rise}}{\text{run}}$ and simplify.

MATCHED EXAMPLE 4 | Finding the slope from two points

Calculate the slope of the line for each of the following.

a A straight line through the points (3, 8) and (5, 14).

b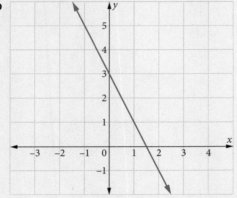

c A straight line drawn from the following table of values.

X	−2	−1	0	1	2
Y	−6	−3	1	4	7

Steps	Working
a Use slope $= \dfrac{y_2 - y_1}{x_2 - x_1}$ for (x_1, y_1) and (x_2, y_2) and simplify.	
b 1 Select two points on the line that can be clearly read from the graph. 2 Use slope $= \dfrac{y_2 - y_1}{x_2 - x_1}$ for (x_1, y_1) and (x_2, y_2) and simplify.	
c 1 Select two points from the table. 2 Use slope $= \dfrac{y_2 - y_1}{x_2 - x_1}$ for (x_1, y_1) and (x_2, y_2) and simplify.	

MATCHED EXAMPLE 5 — Finding the straight line equation from the slope and intercept

Find the equation of each of the following straight lines by finding the slope and y-intercept.

a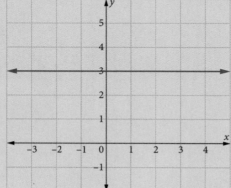

b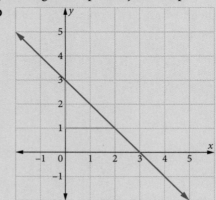

c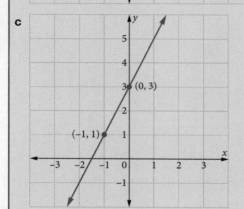

Steps	Working
a 1 Is the line sloping up or down from left to right?	
2 Where does the graph cross the y-axis?	
3 Identify the values for a and b in the equation $y = ax + b$.	
4 Write the equation of the line.	
b 1 Is the line sloping up or down from left to right?	
2 Use slope $= \dfrac{\text{rise}}{\text{run}}$ and simplify.	
3 Where does the graph cross the y-axis?	
4 Identify the values for a and b in the equation $y = ax + b$.	
5 Write the equation of the line.	
c 1 Use slope $= \dfrac{y_2 - y_1}{x_2 - x_1}$ for (x_1, y_1) and (x_2, y_2) and simplify.	
2 Where does the graph cross the y-axis?	
3 Identify the values for a and b in the equation $y = ax + b$.	
4 Write the equation of the line.	

MATCHED EXAMPLE 6 | Working with the constant rate of change and initial value

Adventure World charges its visitors a combination of an entry cost and a rate per ride taken according to the graph shown.

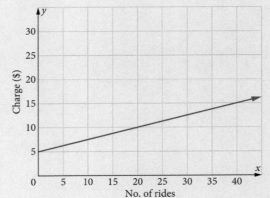

Find the
a vertical intercept of the graph
b slope of the graph
c equation of the graph using C for the total charge and m as the rate per ride.

From the equation, find the
d entry cost
e rate of charge per ride
f total cost of going on 25 rides
g number of rides that have a total cost of $15, rounding your answer to the nearest number of ride.

Steps	Working
a Read from the graph.	
b 1 Select two points on the line that can be clearly read from the graph. 2 Use slope $= \dfrac{y_2 - y_1}{x_2 - x_1}$ for (x_1, y_1) and (x_2, y_2) and simplify.	
c 1 Identify a and b in the equation $y = ax + b$. 2 Write the equation $y = ax + b$ using the variables given.	
d Find the initial value (vertical intercept).	
e Find the constant rate of change (slope).	
f Substitute the value into the equation and solve, using CAS if necessary.	

g 1 Substitute the value into the equation and solve, using CAS if necessary.

2 Write the answer rounding to the nearest number of rides.

MATCHED EXAMPLE 7 — Dealing with the domain of interpretation in real-life problems

Year 11 students are organising a fancy dress party. The total cost for the party will include $650 venue hire, $170 for music and $32 per head for food. The maximum capacity of the venue is 150.

a Find the linear equation in the form $C = an + b$ for the total cost of the party, C, for n students.
b Use the equation to find the total cost of the games if 7 students attended.
c Explain why your answer to part **b** makes no sense in real life.
d Use the equation to find the total cost of the food if 900 students attended.
e Explain why your answer to part **d** makes no sense in real life.

Steps	Working
a 1 Identify the constant rate of change and initial value.	
2 Write the total cost equation.	
b Substitute the value of n into the equation and evaluate. Write the answer in words.	
c What would have happened in real life?	
d Substitute the value of n into the equation and evaluate. Write the answer in words.	
e What would have happened in real life?	

MATCHED EXAMPLE 8 | Modelling profit and loss

Vegetarian pizzas are sold for $11.25 each, and the cost C of making n vegetarian pizzas is given by the equation

$C = 180 + 7n$

a Find the revenue equation in terms of n.
b Find the profit equation in terms of n.
c How much profit would be made if 150 vegetarian pizzas were sold?
d How much profit would be made if 40 vegetarian pizzas were sold?
e How many vegetarian pizzas need to be sold to make at least $1500 profit? Explain why you need to round *up* for this calculation.

Steps	Working
a Use the price of one item to calculate the revenue from selling n items.	
b Use the profit equation and simplify.	
c Substitute $n = 150$ into the profit equation and write the answer.	
d Substitute $n = 40$ into the profit equation and write the answer.	
e 1 Let profit equal 1500 and solve for n, using CAS if necessary. **2** Round the answer according to the question. The answer needs to be a whole number since n represents the number of vegetarian pizzas.	

MATCHED EXAMPLE 9 — Graphing linear relations in the form $Ax + By = C$ by hand

For the equation $3x - 2y = 9$

a find the x-intercept
b find the y-intercept
c and hence sketch the graph by hand.

Steps — **Working**

a Substitute $y = 0$ and solve the equation.
   Write the coordinates of the x-intercept.

b Substitute $x = 0$ and solve the equation.
   Write the coordinates of the y-intercept.

c Sketch the graph on a Cartesian plane by marking in the x- and y-intercepts and drawing a straight line through the two intercepts.

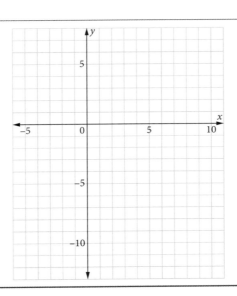

Using CAS 2: Graphing and solving simultaneous equations p. 182

p. 184

MATCHED EXAMPLE 10 | Solving problems using simultaneous equations

Daphne spent $139.20 on pastries and cakes. She purchased a total of 9 items. If the pastries cost $12.60 each and the cakes cost $25.50 each, how many pastries and cakes did Daphne purchase?

Steps	Working
1 Identify the unknowns and assign a pronumeral for each of them.	
2 Set up two equations by converting the given information into mathematical symbols. Convert units where necessary.	
3 Solve the simultaneous equations using CAS. **TI-Nspire**	**ClassPad**
4 Answer the question in sentence form.	

MATCHED EXAMPLE 11 | Interpreting line segment graphs

A large oil container holds 20 litres. The line segment linear graph shows the rate at which the oil container fills with oil.

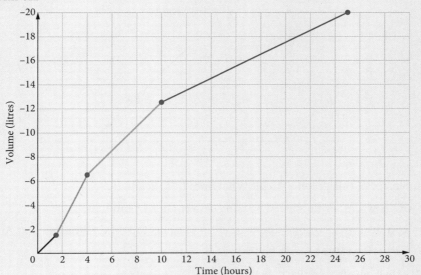

a Explain how we know the oil container fills at four different rates.
b Approximately how many litres does the oil container hold after 10 hours?
c After how many hours from the start of oil machine is the oil container filled to capacity?
d Approximately during which times does the container fill the fastest?
e What is the rate, in litres per hour, that the container is filling in the last 15 hours?

Steps	Working
a Refer to the line segments and their slopes.	
b Read from the graph and note how the vertical scale is written.	
c Read from the graph.	
d Identify which of the line segments has the greatest slope.	
e Find the slope of the last line segment. Convert the rate to litres per hour.	

MATCHED EXAMPLE 12 — Interpreting line segment distance-time graphs

The distance-time graph below shows how Catherine walked to Amy's house, stayed there for a while, then walked back home with Amy.

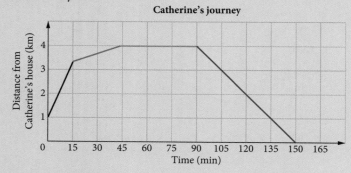
Catherine's journey

a The vertical intercept of this graph is 1 km. What does this mean?
b Explain how we know Catherine started walking slower after 15 minutes.
c Explain how we know the horizontal line segment represents the time Catherine is at Amy's house.
d How long did Catherine stay at Amy's house?
e What is the distance between Catherine's house and Amy's house?
f Calculate Catherine's and Amy's speed in kilometres/hour during the last section of their journey.

Steps	Working
a The vertical intercept is the initial value. What is the vertical scale measuring?	
b Refer to the slope of the line segments, which measures speed.	
c Refer to the vertical axis value.	
d Read from the graph.	
e Read the distance using the vertical scale of the graph.	
f Calculate the slope from the graph in kilometres/minute and convert to kilometres/hour.	

MATCHED EXAMPLE 13 | Interpreting step graphs

This step graph shows the ride fares at a theme park.

a Find the charge for rides for
  i 40 minutes
  ii 5 hours
  iii 7 hours
  iv 9 hours 30 minutes.

b What range of times can a person ride for $20?

c What does the arrow on the $35 step mean?

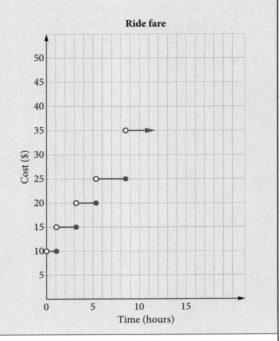

Steps	Working
a i Read from the graph.	
ii Read from the graph.	
iii Read from the graph.	
iv Read from the graph.	

b **1** Find the relevant step on the graph.

Remember, a dot means the value is included and a circle means it's not included.

Write the answer.

c The arrow indicates the line continues.

CHAPTER 5

MATRICES

MATCHED EXAMPLE 1	Understanding the order of matrices

The table shows the number of different items sold in a café on five days of the week.

Item	Days				
	Mon	Tue	Wed	Thu	Fri
Cake	60	50	75	20	90
Coffee	150	120	90	130	100
Ice cream	30	80	60	40	80
Croissant	180	150	130	160	155

p. 208

Find

Steps	Working
a the matrix M that could be used to show this information, stating its order and number of elements.	
Rewrite the information in the table as a matrix.	
b the matrix that could be used to show the number of cups of coffee sold on Wednesdays in the café and state its order.	
Find the information in the table and write as a matrix.	
c the 1×5 matrix that could be used to show the number of cakes sold on the five days of the week.	
Find the information in the table and write as a matrix with the given order.	
d the 5×1 matrix that could be used to show the number of coffees sold on the five days of the week.	
Find the information in the table and write as a matrix with the given order.	

Chapter 5 | Matrices 77

e the 1 × 4 matrix that could be used to show the number of all the items sold on Tuesday.

Find the information in the table and write as a matrix with the given order.

f the 4 × 1 matrix that could be used to show the total number of items sold on all the five days of the week

Find the information in the table and write as a matrix with the given order.

g Copy the following labelled matrix showing the information from the table and fill in the missing numbers.

$$\begin{array}{c} \\ \text{Monday} \\ \text{Tuesday} \\ \text{Wednesday} \\ \text{Thursday} \\ \text{Friday} \end{array} \begin{array}{cccc} \text{Cake} & \text{Coffee} & \text{Ice cream} & \text{Croissant} \\ \left[\begin{array}{cccc} 60 & \square & \square & \square \\ \square & 120 & \square & \square \\ \square & \square & \square & 130 \\ 20 & \square & \square & \square \\ \square & \square & \square & \square \end{array} \right] \end{array}$$

Find the information in the table and complete the matrix.

$$\begin{array}{c} \\ \text{Monday} \\ \text{Tuesday} \\ \text{Wednesday} \\ \text{Thursday} \\ \text{Friday} \end{array} \begin{array}{cccc} \text{Cake} & \text{Coffee} & \text{Ice cream} & \text{Croissant} \\ \left[\begin{array}{cccc} 60 & & & \\ & 120 & & \\ & & & 130 \\ 20 & & & \\ & & & \end{array} \right] \end{array}$$

MATCHED EXAMPLE 2 Identifying types of matrices

For each of the following matrices state the order and whether it's a row, column, square, zero or identity matrix.

a $\begin{bmatrix} 1 & 0 & 0 & 0 \\ 0 & 1 & 0 & 0 \\ 0 & 0 & 1 & 0 \\ 0 & 1 & 0 & 0 \end{bmatrix}$
b $\begin{bmatrix} 1 & 2 & 3 & 4 \end{bmatrix}$
c $\begin{bmatrix} 1 & 0 & 0 \\ 0 & 1 & 0 \\ 0 & 0 & 1 \end{bmatrix}$

d $\begin{bmatrix} 0 & 0 & 0 & 0 & 0 \\ 0 & 0 & 0 & 0 & 0 \\ 0 & 0 & 0 & 0 & 0 \\ 0 & 0 & 0 & 0 & 0 \\ 0 & 0 & 0 & 0 & 0 \end{bmatrix}$
e $\begin{bmatrix} 3 \\ 3 \\ 3 \\ 3 \end{bmatrix}$

Steps	Working
Does the matrix have	
• just one row or just one column	
• the same number of rows and columns	
• all zeros	
• 1s it in the diagonal from top left to bottom right and zeros everywhere else?	

MATCHED EXAMPLE 3 — Adding, subtracting and multiplying matrices by a scalar

If $A = \begin{bmatrix} 2 & 2 \\ 5 & 0 \\ 1 & 1 \end{bmatrix}$, $B = \begin{bmatrix} 3 & -2 & 4 & 1 \end{bmatrix}$, $C = \begin{bmatrix} 1 & -1 & 0 & 1 \end{bmatrix}$ and $D = \begin{bmatrix} 1 & 2 \\ -1 & -2 \\ 1 & 2 \end{bmatrix}$,

calculate the following, giving a reason if the addition or subtraction is not defined.

a $D - A$ **b** $B + C$ **c** $2B$ **d** $\frac{1}{2}D$ **e** $A - 3D$ **f** $2B + A$

Steps — **Working**

1 Check that the matrices have the same order.

2 Add or subtract corresponding elements.

3 Multiply each element by the scalar.

MATCHED EXAMPLE 4 — Finding missing elements in matrix equations

Find the value of x, y and z in each of the following.

a $\begin{bmatrix} 5x & 1 \\ 0 & -3 \end{bmatrix} - \begin{bmatrix} 2 & -3 \\ 2y & 4 \end{bmatrix} = \begin{bmatrix} 8 & 4 \\ -2 & z \end{bmatrix}$

b $2\begin{bmatrix} 2 \\ x \\ 1 \end{bmatrix} + \begin{bmatrix} y \\ 4 \\ 10 \end{bmatrix} = 3\begin{bmatrix} 6 \\ 6 \\ z \end{bmatrix}$

Steps	Working
a Using the elements in the same row and column of each matrix, write down equations involving the unknowns. Solve, using CAS if necessary.	
b Using the elements in the same row and column of each matrix, write down equations involving the unknowns. Solve, using CAS if necessary.	

MATCHED EXAMPLE 5: Working with matrices using addition, subtraction and scalar multiplication

The cost prices of four different guitars in a store are $6799, $8899, $9999 and $10 980. The selling price of each of these four guitars is 2.2 times the cost price.

Steps	Working
a Show a matrix calculation involving a column matrix that will give the selling price of each guitar.	
Use scalar multiplication.	
b If the selling price also needs to allow for a $85 commission for the salesman, show how this can be included in the matrix calculation.	
Use matrix addition.	
c The store has a sale where guitars with a cost price less than $10 000 have their selling price reduced by $500, and guitars with a cost price greater than $9000 have their selling price reduced by $1000. Show how this can be included in the matrix calculation of the selling price.	
Use matrix subtraction.	

MATCHED EXAMPLE 6 | Multiplying matrices

If $A = \begin{bmatrix} 1 & 5 \\ 4 & 2 \\ 3 & 0 \end{bmatrix}$, $B = \begin{bmatrix} 2 & 1 \\ 1 & 2 \end{bmatrix}$, $C = \begin{bmatrix} 2 \\ 5 \\ 8 \end{bmatrix}$ and $D = \begin{bmatrix} 3 & 1 & 1 \end{bmatrix}$, for each of the following

a BC **b** AB **c** CD **d** DC
e C^2 **f** $B^2 + 3B$ **g** CA

  i state whether or not the expression is defined, giving a reason.

For those that are defined

  ii state the order of the answer before performing the calculation
  iii do the calculation to find the answer.

Steps	Working
a **i** Do the number of columns in B equal the number of rows in C?	
b **i** Do the number of columns in A = number of rows in B? **ii** How many rows does A have? How many columns does B have? **iii** Calculate AB.	
c **i** Do the number of columns in C = number of rows in D? **ii** How many rows does C have? How many columns does D have? **iii** Calculate CD.	

d **i** Do the number of columns in *D* equal the number of rows in *C*?

   ii How many rows does *D* have?
   How many columns does *C* have?

   iii Calculate *DC*.

e **i** Is *C* a square matrix?

f **i** Is *B* a square matrix?

   ii How many rows does *B* have?
   How many columns does *B* have?
   Is the sum possible?

   iii Calculate $B^2 + 3B$.

g **i** Do the number of columns in *C* equal the number of rows in *A*?

MATCHED EXAMPLE 7 Finding the determinant and inverse of a 2 × 2 matrix

For each of the following matrices, find
 i the determinant (if it exists)
 ii the inverse (if it exists).

a $A = \begin{bmatrix} 1 & 4 \\ 3 & 2 \end{bmatrix}$ **b** $B = \begin{bmatrix} 3 & 3 \\ 1 & 1 \end{bmatrix}$ **c** $C = \begin{bmatrix} 1 & 4 & 4 \\ 2 & 2 & 5 \end{bmatrix}$ **d** $D = \begin{bmatrix} 7 & 10 \\ 2 & 3 \end{bmatrix}$

Steps	Working
a i Is it a square matrix? For $\begin{bmatrix} a & b \\ c & d \end{bmatrix}$, use determinant $= ad - bc$.	
ii Is the determinant equal to 0? If not, use inverse $= \dfrac{1}{ad-bc}\begin{bmatrix} d & -b \\ -c & a \end{bmatrix}$	
b i Is it a square matrix? For $\begin{bmatrix} a & b \\ c & d \end{bmatrix}$, use determinant $= ad - bc$.	
ii Is the determinant equal to 0? If not, use inverse $= \dfrac{1}{ad-bc}\begin{bmatrix} d & -b \\ -c & a \end{bmatrix}$	
c i Is it a square matrix? For $\begin{bmatrix} a & b \\ c & d \end{bmatrix}$, use determinant $= ad - bc$.	
ii Is the determinant equal to 0? If not, use inverse $= \dfrac{1}{ad-bc}\begin{bmatrix} d & -b \\ -c & a \end{bmatrix}$	
d i Is it a square matrix? For $\begin{bmatrix} a & b \\ c & d \end{bmatrix}$, use determinant $= ad - bc$.	
ii Is the determinant equal to 0? If not, use inverse $= \dfrac{1}{ad-bc}\begin{bmatrix} d & -b \\ -c & a \end{bmatrix}$.	

MATCHED EXAMPLE 8 Solving simultaneous equations with two unknowns using matrices

Solve the following simultaneous equations using matrices, showing all the steps.

$4x - 2y = -2$
$3x + 2y = 9$

Steps	Working
1 Write the simultaneous equations in matrix form.	
2 Find the determinant of the matrix.	
3 Find the inverse of the square matrix in the matrix equation.	
4 Multiply on the left by the inverse on both sides of the matrix equation and simplify. A matrix multiplied by its inverse is the identity matrix. A matrix multiplied by the identity matrix is unchanged.	
5 Multiply the matrices on the right of the equation. Equate the elements to solve the simultaneous equations.	

MATCHED EXAMPLE 9 | Solving problems using inverse matrices

A factory has three identical machines which assemble three different models of a smart phone (A, B and C).

Machine 1 assembles three model As, two model Bs and a model C in 464 minutes.

Machine 2 assembles two model As, a model B and a model C in 302 minutes.

Machine 3 assembles two model As, three model Bs and four model Cs in 666 minutes.

Let a = the amount of time in minutes it takes for a machine to assemble one of model A.

Let b = the amount of time in minutes it takes for a machine to assemble one of model B.

Let c = the amount of time in minutes it takes for a machine to assemble one of model C.

a Write three simultaneous equations in terms of a, b and c.

b Write the simultaneous equations in matrix form.

c Solve the matrix equation and hence, find how long it takes a machine to assemble each of the three computer models.

Steps	Working
a Use the information in the question to write three simultaneous equations. Write the pronumerals so they line up under each other.	
b Rewrite in matrix form, adding zeros where necessary.	
c 1 Solve with CAS by finding the inverse of the 3 × 3 matrix and multiplying on the left by the inverse on both sides of the matrix equation. 2 Write the answer.	

MATCHED EXAMPLE 10 — Solving costing and pricing problems using matrices

The manager of a clothing store purchases short sleeve T-shirts for $20 each and long sleeve T-shirts for $23. In the last two months, he has purchased the number of T-shirts shown.

	Short sleeve T-shirts	Long sleeve T-shirts
Month 1	35	20
Month 2	42	31

Steps	Working
a Find the two matrices that can be multiplied to give the total purchase cost of T-shirts in each of the two months and complete the multiplication. State the total T-shirts costs for each month.	
We need a matrix product that calculates number of short sleeves × cost of short sleeves + number of long sleeves × cost of long sleeves.	
b The manager of a clothing store sells goods at 120% of the cost price. He recorded his purchase costs over the last two months for T-shirts and three other types of clothing in the following table. 		Week 1	Week 2	 |---------|--------|--------| 	T-shirts			 	Jeans	$3672	$4250	 	Sweaters	$4130	$3280	 	Skirts	$3330	$2050	 **i** Represent these costs in a 4 × 2 cost matrix, C. **ii** Using scalar multiplication, represent the selling prices of these goods in a 4 × 2 matrix, S.	
i The table already has 4 rows and 2 columns. Fill in the missing information from part a.	
ii Convert the percentage to a decimal and multiply the cost matrix by the decimal.	

c **i** Create a profit matrix.
  ii Calculate the total profit to be made if all goods purchased over these two months are sold.

  i To create a profit matrix:

  profit = selling price − cost price

  ii The total profit can be found by adding all of the elements in the profit matrix.

MATCHED EXAMPLE 11 — Working with communication matrices and diagrams

Steps | **Working**

a The communication matrix M shows how direct messages can be sent between four people: Anna (A), Beth (B), Cho (C) and David (D).

$$M = \text{sender} \begin{array}{c} \\ A \\ B \\ C \\ D \end{array} \overset{\text{receiver}}{\begin{bmatrix} A & B & C & D \\ 0 & 1 & 0 & 1 \\ 1 & 0 & 0 & 1 \\ 0 & 1 & 0 & 1 \\ 0 & 1 & 1 & 0 \end{bmatrix}}$$

i List who each person can send direct messages to.
ii Explain why the diagonal from the top left to the bottom right is all zeros.
iii Draw a communication diagram showing the communication links given in the matrix.
iv How could Cho get a message to Anna in two steps?

i Look at each row in order.
A '1' means the person can send a direct message.

ii Refer to redundant links.

iii Draw a diagram with arrows that match the list of possible direct messages.

iv Find how the message could be passed on in two steps.

b Write the communication matrix for the following communication diagram.

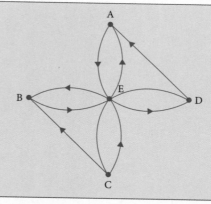

▶ Set up a 5 × 5 square matrix where a '1' indicates communication and a '0' indicates non-communication.

MATCHED EXAMPLE 12 | Working with two-step communication

For the communication matrix representing the connections between four computers, find the following.

$$M = \text{from} \begin{array}{c} \\ A \\ B \\ C \\ D \end{array} \begin{array}{c} \text{to} \\ \begin{bmatrix} A & B & C & D \\ 0 & 0 & 1 & 1 \\ 1 & 0 & 0 & 1 \\ 1 & 0 & 0 & 0 \\ 0 & 0 & 1 & 0 \end{bmatrix} \end{array}$$

a The number of ways B can connect with D by connecting directly to one other computer.
b The list of all the two-step connections from D to A.
c The total number of redundant two-step connections.
d The list of redundant two-step connections from A to A.

Steps	Working
a Find M^2 using CAS and read the number of two-step connections from the matrix.	
b Use M to find the two-step connections.	
c Sum the values in the top left to bottom right diagonal of M^2.	
d Use M^2 to find the number of redundant two-step connections. Use M to find the redundant two-step connections.	

MATCHED EXAMPLE 13 — Constructing transition matrices

Construct transition matrices for each of the following.

Steps	Working

a For this transition diagram showing changes from one year to the next
  i find x, y, z
  ii construct the matching transition matrix.
	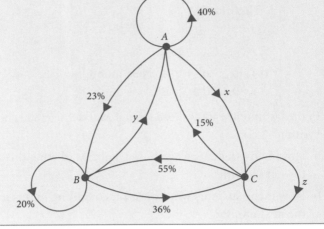

i All the arrow percentages *from* a single point add up to 100%. Solve for the unknowns, using CAS if necessary.	
ii Convert all the percentages to decimals and construct the matrix.	

b If a roulette player wins in the current game, there is a 30% chance he wins in the next game. If he loses in the current game, then there is a 78% chance he might lose in the next.

1 Set up a 2×2 matrix using W for 'win' and L for 'lose' with a transition from 'Current game' to 'Next game'. Enter the percentages from the question as decimals.	
2 Enter the remaining elements of the matrix by using the fact that the columns of a transition matrix must add up to 1.	

MATCHED EXAMPLE 14 — Interpreting transition matrices

An online learning website offers three skill development courses for students to expand their job opportunities: Digital Marketing (D), Video Editing (V) and Business Writing (B). Each month, the students choose one of the courses. The transition matrix shows how the students' choices change from month to month.

$$T = \begin{bmatrix} 0.6 & 0.4 & 0.3 \\ 0.3 & 0.2 & 0.5 \\ 0.1 & 0.4 & 0.2 \end{bmatrix} \begin{matrix} D \\ V \\ B \end{matrix} \text{ Next month}$$

This month: $D \quad V \quad B$

In the last month, 4000 chose Digital Marketing, 3300 chose Video Editing and 5800 chose Business Writing.

a How many students who chose Digital Marketing this month are expected to choose Video Editing in the next month?

b How many students who chose Video Editing in this month are expected to choose Video Editing in the next month?

c How many students are expected to choose Business Writing in the next month?

d If 52% of the students chose Digital Marketing one month, what percentage are expected to choose Video Editing in the next month?

Steps	Working
a Locate the relevant element in the transition matrix and multiply by the number of students.	
b Locate the relevant element in the transition matrix and multiply by the number of students.	
c Locate the relevant elements in the transition matrix, multiply by the number of students in each case and add.	
d Locate the relevant element in the transition matrix and multiply by the percentage.	

MATCHED EXAMPLE 15 — Finding the steady-state matrix

A fleet of ships starts at either one of the two seaports, R or S. By the end of the month, the ships end up at one of the ports according to the transition matrix T.

$$T = \begin{bmatrix} 0.65 & 0.15 \\ 0.35 & 0.85 \end{bmatrix} \begin{matrix} R \\ S \end{matrix} \quad \text{Next month}$$

This month: R, S

At the start of a particular month, there are 1200 ships at seaport R and 980 ships at seaport S.

Steps	Working
a Find the steady-state matrix.	
1 Use CAS and the rule $S_n = T^n S_0$ for two large consecutive values of n.	
2 Are the two state matrices the same?	
b How many ships will be at each seaport in the long term?	
Read from the steady-state matrix.	
c What percentage of ships are at seaport S in the long term? Round your answer to the nearest percentage.	
Calculate from the steady-state matrix.	

CHAPTER 6
RELATIONSHIPS BETWEEN NUMERICAL VARIABLES

p. 271

MATCHED EXAMPLE 1 Identifying explanatory and response variables

For each of the following identify the explanatory variable, giving a reason for your answer.
a Can the amount of rainfall be explained by the number of trees cut?
b What is the relationship between the age of a person and the hours they spend watching TV?
c Can the number of hours a student spends studying be used to predict their marks?

Steps	Working
a 1 Identify the two variables. 2 Do the words 'explain changes' or 'predict' appear in the question? If not, which variable is most likely to affect the other?	
b 1 Identify the two variables. 2 Do the words 'explain changes' or 'predict' appear in the question? If not, which variable is most likely to affect the other?	
c 1 Identify the two variables. 2 Do the words 'explain changes' or 'predict' appear in the question? If not, which variable is most likely to affect the other?	

Using CAS 1:
Constructing a scatterplot
p. 274

MATCHED EXAMPLE 2 | Interpreting scatterplots

A study was conducted on the age of a person and the number of pets they own and a scatterplot was plotted of the data.

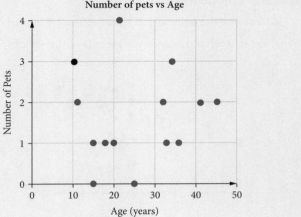

Steps	Working
a What is the explanatory variable?	
The explanatory variable appears on the x-axis.	
b What is the response variable?	
The response variable appears on the y-axis.	
c How many people were in the study?	
Count the number of dots.	
d What does the black dot represent?	
Read from both axes.	
e How many teenagers were in the study?	
Read from the x-axis and count the dots between 13 and 20.	
f How many pets does the oldest person in the study have?	
Read from the y-axis.	

MATCHED EXAMPLE 3 Scatterplots and association

For each of the following scatterplots
  i describe the association between the two variables in terms of direction, form and strength
  ii explain what this means in terms of the variables.

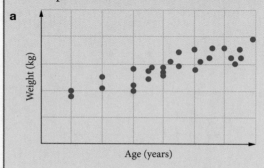

a (Weight (kg) vs Age (years))
b (Exam scores vs Time spent watching TV (hours))
c (Height (ft) vs No. of T-shirts owned)
d (No. of wristwatches owned vs Salary)

Steps	Working
a i Is the data sloping up or down? Does the data follow a straight line pattern? How spread out are the data points? ii Refer to the variables.	
b i Is the data sloping up or down? Does the data follow a straight line pattern? How spread out are the data points? ii Refer to the variables.	
c i Is the data sloping up or down? Does the data follow a straight line pattern? How spread out are the data points? ii Refer to the variables.	
d i Is the data sloping up or down? Does the data follow a straight line pattern? How spread out are the data points? ii Refer to the variables.	

MATCHED EXAMPLE 4 Exploring causation

For each of the following associations between pairs of variables, state another variable that could be the underlying cause of the association between the two.

Steps	Working
a A negative association between *ice cream sales* and *hot chocolate sales*.	
Which variable might be causing changes in both?	
b A positive correlation between *the number of hours a mobile is charged* and *the number of hours spent using the phone*	
Which variable might be causing changes in both?	
c A positive association between the *number of cars owned per household* and the *amount of money spent on designer clothes*	
Which variable might be causing changes in both?	

MATCHED EXAMPLE 5 — Rounding to decimal places versus significant figures

Round each number to
  i two decimal places
  ii two significant figures

a 16.799　　b 40 444　　c 12.329　　d 0.1652
e 12 136.356　　f 6.525　　g 3.143

Steps	Working
a i Focus on the first two decimal places. ii Focus on the first two significant figures.	
b i Focus on the first two decimal places. ii Focus on the first two significant figures.	
c i Focus on the first two decimal places. ii Focus on the first two significant figures.	
d i Focus on the first two decimal places. ii Focus on the first two significant figures.	
e i Focus on the first two decimal places. ii Focus on the first two significant figures.	
f i Focus on the first two decimal places. ii Focus on the first two significant figures.	
g i Focus on the first two decimal places. ii Focus on the first two significant figures.	

MATCHED EXAMPLE 6 — Working with the line of good fit equation

A line of good fit has been drawn on the following scatterplot showing the time spent reading by students of various ages.

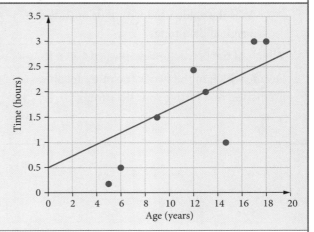

Steps	Working
a Find the intercept of the line.	
Where does the line touch the vertical axis?	
b Calculate the slope of the line, correct to two significant figures.	
1 Choose two easy-to-identify points on the line. 2 Calculate the slope from these two points, giving the answer to the required rounding.	
c What is the equation of the line of good fit?	
Use *response variable* = $a + b \times$ *explanatory variable*, where a is the intercept and b is the slope.	
d i Use the line to predict the number of minutes an 8-year-old would spend reading, rounding to two significant figures.	
Substitute the value into the equation in place of *age* and solve for *time*, giving the answer to the required rounding.	
ii State whether this prediction involves interpolation or extrapolation, giving a reason.	
Was the value used within or outside the original data range?	
e i Use the line to predict the number of minutes a 4-year-old would spend reading, rounding to two significant figures.	
Substitute the value into the equation in place of *age* and solve for *time*, giving the answer to the required rounding.	

	ii State whether this prediction involves interpolation or extrapolation, giving a reason.
	Was the value used within or outside the original data range?
f	Which of the predictions in parts **d** and **e** is more reliable? Justify your answer.
	Decide which of the two is the more reliable prediction and justify your decision.

MATCHED EXAMPLE 7 | Interpreting a line of good fit equation

A study of the association between job experience and average weekly wages has resulted in the following line of good fit equation

wages = 890 + 8.8 × *experience*

where wages are measured in dollars and experience in years.

a Identify and interpret the intercept. **b** Identify and interpret the slope.

Steps	Working
For the equation of the line of good fit *response variable* = $a + b \times$ *explanatory variable* a = the intercept of the line b = the slope of the line	**a**
	b

CHAPTER 7
GRAPHS AND NETWORKS

MATCHED EXAMPLE 1 | Identifying isomorphic graphs

For each of the following pairs of graphs, state whether or not they are isomorphic and give a reason for your answer.

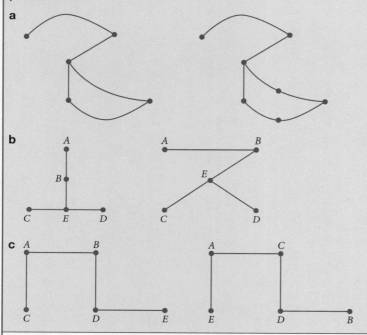

Steps	Working
Do the graphs have the same number of vertices?	
Do the graphs have the same number of edges?	
Do the graphs show *exactly* the same connections?	

MATCHED EXAMPLE 2 — Identifying the features of graphs

For each of the following graphs

    i count and list the vertices, edges and faces

    ii show that the degree sum is twice the number of edges.

a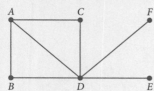

b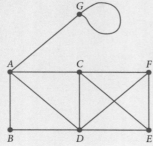

Steps	**Working**
a **i** Count and list the number of vertices and edges. Count and list the number of enclosed regions plus the region outside the graph. **ii** Find the degree of each vertex and sum them. Show that multiplying the number of edges by 2 gives the same result.	 	Vertex	A	B	C	D	E	F	Sum
---	---	---	---	---	---	---	---
Degree								
b **i** **1** Redraw the graph to uncross the intersecting edges that have no vertex at the point of intersection. **2** Count and list the number of vertices and edges. Count and list the number of enclosed regions plus the region outside the graph. **ii** Find the degree of each vertex and sum them. Show that multiplying the number of edges by 2 gives the same result.	 | Vertex	A	B	C	D	E	F	G	Sum
---	---	---	---	---	---	---	---	---
Degree									

MATCHED EXAMPLE 3 Finding adjacency matrices

Represent the following graph using an adjacency matrix.

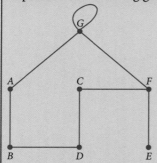

Steps **Working**

1. Set up the square matrix with the vertices as row and column labels.

   Fill in the first row by counting the connections between A and each of the other vertices.

$$\begin{array}{c} \\ A \\ B \\ C \\ D \\ E \\ F \\ G \end{array} \begin{array}{cccccccc} A & B & C & D & E & F & G \\ \left[\begin{array}{ccccccc} & & & & & & \\ & & & & & & \\ & & & & & & \\ & & & & & & \\ & & & & & & \\ & & & & & & \\ & & & & & & \end{array} \right] \end{array}$$

2. Fill in the first column by copying the first row.

   Write you answer in the above matrix.

3. Continue the steps for each row and column.

   Write you answer in the above matrix.

MATCHED EXAMPLE 4 Verifying Euler's formula

For the graph shown

a redraw it to show it is a planar graph

b state whether or not it is a connected graph, giving a reason

c verify that Euler's formula works or show that it doesn't.

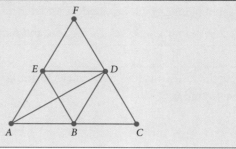

Steps	Working
a To uncross the edges, move edge AD around the outside of the graph.	
b Use the definition of a connected graph.	
c Count the number of vertices, faces and edges, and substitute into **Euler's** formula to see if the result is 2.	

MATCHED EXAMPLE 5 | Using Euler's formula

a A connected planar graph has 7 edges and 4 faces. How many vertices does it have?

b A connected planar graph has 15 vertices and 6 faces. How many edges does this graph have?

Steps	Working
a Substitute the known values into Euler's formula and solve to find v.	
b Substitute the known values into Euler's formula and solve to find e.	

MATCHED EXAMPLE 6 | Identifying subgraphs

For each of the following graphs, state whether it is a subgraph of

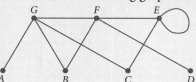

giving reasons for your answers.

a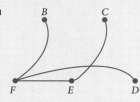

b

c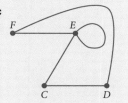

d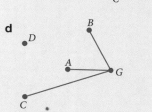

Steps	Working
Does the graph contain *only* vertices and edges from the original graph?	

MATCHED EXAMPLE 7 — Classifying walks shown on a graph

For each of the following walks, state whether it is a trail, path, circuit, cycle or a walk only and give a reason for your answer.

a 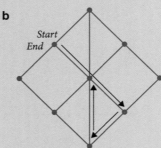 b c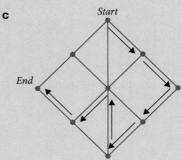

d 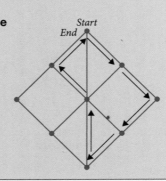 e

Steps	Working
Use the Walk classification chart to ask three questions, in this order, for each one: 1 Does the walk have repeated edges? 2 Does the walk have repeated vertices? 3 Does the walk start and finish at the same vertex?	

MATCHED EXAMPLE 8 — Classifying walks from a list of vertices

Anika is jogging along paths in park. For each of the following walks, state whether it is a trail, path, circuit, cycle or a walk only and give a reason for your answer.

a P-Q-R-V-S-U-T
b U-S-V-R-Q-P-U
c R-V-S-U-V-Q-R
d T-U-V-Q-R-V-S
e P-Q-V-S-U-V-Q-P
f Q-R-V-U-P-Q

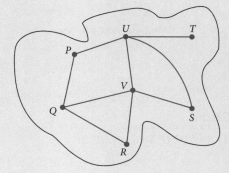

Steps	Working
Use the Walk classification chart to ask three questions, in this order, for each one: 1 Does the walk have repeated edges? 2 Does the walk have repeated vertices? 3 Does the walk start and finish at the same vertex?	

MATCHED EXAMPLE 9 | Finding the shortest path

The network shows the travel times, in hours, from one city to another. Find the shortest time, in hours, that it takes to travel from city A to city B by listing all the options.

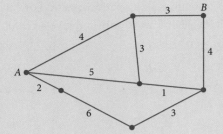

Steps | **Working**

1 Add labels to the vertices.

2 List the path options and calculate the total time of each option.

3 Write the answer.

MATCHED EXAMPLE 10 | Identifying spanning trees

Which of the following graphs are a spanning tree of the graph shown? For those that are spanning trees, verify that the number of edges is one less than the number of vertices. For those that aren't spanning trees, give a reason.

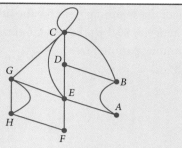

a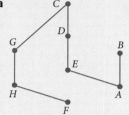

b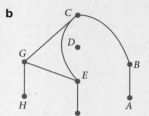

c

d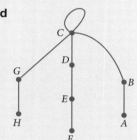

e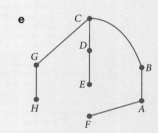

f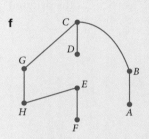

Steps	Working
Is it connected?	
Does it have all the vertices of the original graph?	
Does it have no loops?	
Does it have no multiple edges?	
Does it have no cycles?	

MATCHED EXAMPLE 11 — Finding minimum spanning trees by inspection

Find all the spanning trees for the network shown, and hence, find the total weight of the minimum spanning tree.

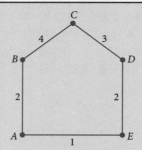

Steps	Working
1 Work out how many edges need to be removed from the graph to create a spanning tree: • The number of edges in a tree is always one less than the number of vertices. • A spanning tree has the same number of vertices as the original graph.	
2 Remove each edge in turn to see which options result in a spanning tree. Calculate the total weight of each spanning tree and find the one with the smallest total weight.	

MATCHED EXAMPLE 12 Finding minimum spanning trees using Prim's algorithm

Use Prim's algorithm to find the minimum spanning tree for the weighted graph shown, and hence, find the total weight of the minimum spanning tree.

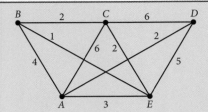

Steps	Working
1 Start at any vertex and choose the edge with the lowest weight connected to this vertex.	
2 Look at *all* the edges connecting to the vertices you've chosen so far (*not just the last vertex connected*) and choose the edge with the lowest weight that doesn't connect to a vertex already in the tree. If there are edges with equal lowest weights, choose one of them.	
3 Repeat step 2 until all the vertices in the graph are included in the tree.	

VARIATION

CHAPTER 8

MATCHED EXAMPLE 1 | Working with direct variation

The cost of flour, c ($), varies directly with their total mass, m (kg), as shown in the table.

m (kg)	0.5	0.75	1.5	1.75
c ($)			2.94	

a Write an equation showing this variation including k, the constant of variation.
b Find the value of k.
c Copy and complete the table.
d Draw a graph of the variation, labelling the values from the table.
e Show a calculation from the table of values that verifies that doubling the mass doubles the cost.
f Find the cost of 15 kg of flour.

Steps	Working
a Write the equation in terms of the given variables and the constant of variation, k.	
b Substitute a pair of known values into the equation and solve for k, using CAS if necessary.	
c Write the equation using the value of k and substitute the values from the table into the equation to complete the table.	

m (kg)	0.5	0.75	1.5	1.75
c ($)			2.94	

d Plot the points from the table.

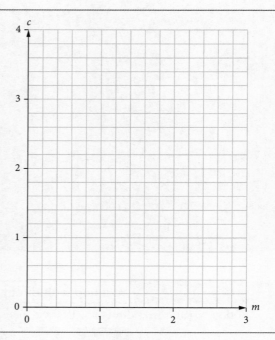

e Use the simplest example from the table for the calculation.

f Substitute the value into the variation equation and solve, using CAS if necessary.

MATCHED EXAMPLE 2 — Working with inverse variation

For a team of construction workers, the number of workers, n, varies inversely with the time, t (days), it takes to build a fence, as shown in the table.

n	8	16	24	36
t (days)				12

a Write an equation showing this variation including k, the constant of variation.
b Find the value of k.
c Copy and complete the table.
d Draw a graph of the variation, labelling the values from the table.
e Show a calculation from the table of values that verifies that doubling the number of workers halves the amount of time it takes to build the fence.
f Find how many days it would take 72 workers to build the fence.

Steps	Working
a Write the equation in terms of the given variables and the constant of variation, k.	
b Substitute a pair of known values into the equation and solve for k, using CAS if necessary.	
c Write the equation using the value of k and substitute the values from the table into the equation to complete the table.	$\begin{array}{\|c\|c\|c\|c\|c\|} \hline n & 8 & 16 & 24 & 36 \\ \hline t\text{ (days)} & & & & 12 \\ \hline \end{array}$
d Plot and label the points from the table.	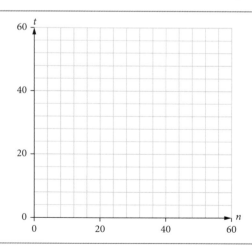
e Use the simplest example from the table for the calculation.	

▶ **f** Substitute the value into the variation equation and solve, using CAS if necessary.

MATCHED EXAMPLE 3 — Identifying direct and inverse variation from a graph

For each of the following, state whether direct, inverse or neither variation is involved, giving a reason for your answer, and find the equation of variation if the variation is direct or inverse.

a
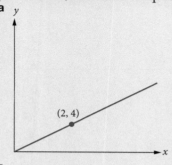

b
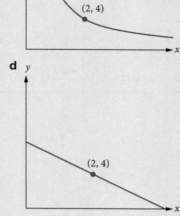

c
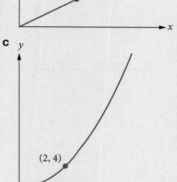

d

Steps	**Working**
a 1 Is the graph a straight line? Does the line go through (0, 0)? Does the graph have the shape x? **2** Using either $y = kx$ or $y = \dfrac{k}{x}$, substitute a point in and solve to find k, using CAS if necessary.	
b 1 Is the graph a straight line? Does the line go through (0, 0)? Does the graph have the shape x? **2** Using either $y = kx$ or $y = \dfrac{k}{x}$, substitute a point in and solve to find k, using CAS if necessary.	

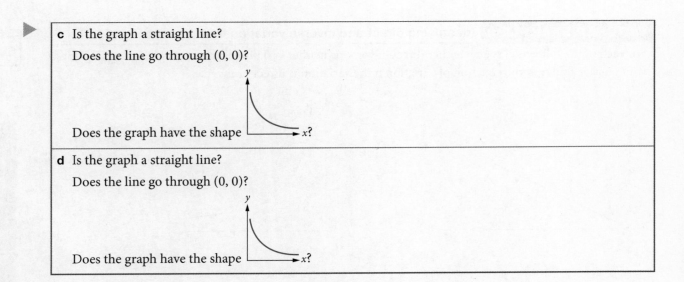

MATCHED EXAMPLE 4 Identifying direct and inverse variation from a graph without (0, 0)

For the following variation graphs
  i describe what is unusual about the horizontal axis
  ii show a calculation that verifies the type of variation
  iii find the variation equation.

a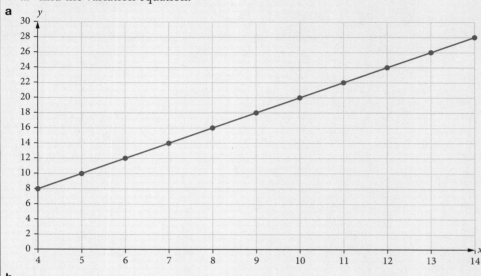

b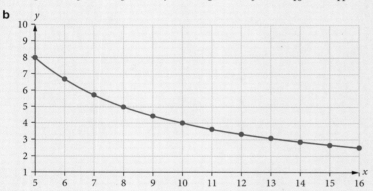

Steps	Working
a i What is the first value on the horizontal axis?	
ii Find two points that can be read from the graph where one x value is double the other x value. Find what happens to the value.	
iii Using either $y = kx$ or $y = \dfrac{k}{x}$, substitute a point in and solve to find k, using CAS if necessary.	
b i What is the first value on the horizontal axis?	
ii Find two points that can be read from the graph where one x value is double the other x value. Find what happens to the value.	

iii Using either $y = kx$ or $y = \dfrac{k}{x}$, substitute a point in and solve to find k, using CAS if necessary.

MATCHED EXAMPLE 5 | Linearising data

Linearise each of the following by transforming the variables as shown.
  i Set up a table of values for the points marked on the graph.
  ii Add a row to the table and include the values for the transformed variable.
  iii Draw the transformed graph with the transformed points to verify it is a straight line.

a Linearise by plotting y against x^2.

b Linearise by plotting y against $\dfrac{1}{x}$.

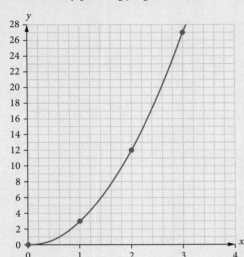
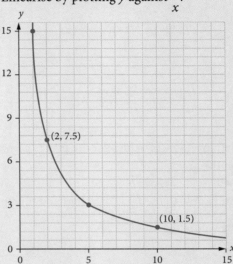

c Linearise by plotting y against $\log x$.

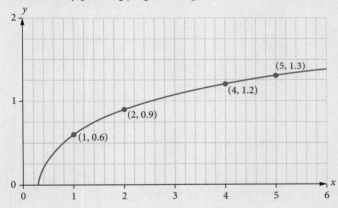

Steps	**Working**
a i Set up a table of values for the points shown on the graph.	
ii Add a row to the table of values and include the values for x^2.	

iii Sketch the linearised graph by plotting the y values against the x^2 values and labelling the horizontal axis x^2. Show the points on the line.

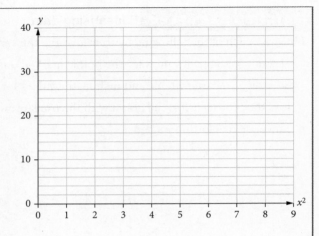

b i Set up a table of values for the points shown on the graph.

   ii Add a row to the table of values and include the values for $\frac{1}{x}$.

   iii Sketch the linearised graph by plotting the y values against the $\frac{1}{x}$ values and labelling the horizontal axis $\frac{1}{x}$. Show the points on the line.

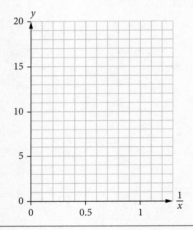

c i Set up a table of values for the points shown on the graph.

   ii Add a row to the table of values and include the values for $\log x$.

   iii Sketch the linearised graph by plotting the y values against the $\log x$ values and labelling the horizontal axis $\log x$. Show the points on the line.

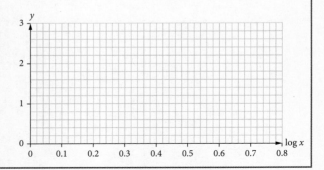

MATCHED EXAMPLE 6 | Modelling non-linear data with curves

For each of the following non-linear scatterplots, list the form of the squared, log and inverse curve of good fit equations that model the data.

a
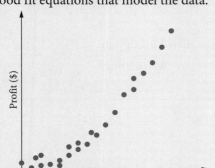

b
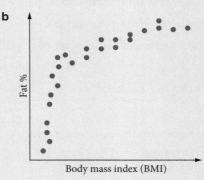

Steps	Working
a 1 Identify the options for the type of relationship. 2 Choose the correct form of the equation from $y = kx^2 + c$ $y = k \log x + c$ $y = \dfrac{k}{x} + c$ using the variable names.	
b 1 Identify the options for the type of relationship. 2 Choose the correct form of the equation from $y = kx^2 + c$ $y = k \log x + c$ $y = \dfrac{k}{x} + c$ using the variable names.	

MATCHED EXAMPLE 7 — Working with non-linear data

The data for the association between time (hours) and the number of new users of a social networking app has been modelled by the inverse relationship given by the following equation.

number of new users = $41 \log(\text{time}) + 50$

a Use the equation to predict the number of new users of the social networking app in

　　i   2 hours　　　　　ii   5 hours　　　　　iii   10 hours

b What shape is the original data?

Steps	Working
a Substitute the value into the equation. Round the answer if necessary.	i ii iii
b 1 Which relationship has been used? 　 2 Which data shape matches this relationship? Decide whether y increases or decreases as x increases.	

CHAPTER 9
MEASUREMENT, SCALE AND SIMILARITY

MATCHED EXAMPLE 1 | **Converting units of measurement**

Convert each of the following units of measurement.

a 1200 square centimetres to square metres

b 0.000135 cubic metres to cubic millimetres

c 0.01253 kilolitres to litres

d 0.5623 km to mm

p. 384

Steps	Working

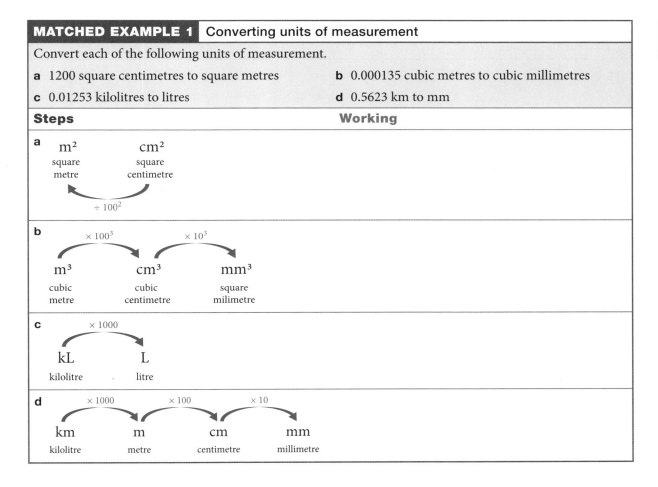

MATCHED EXAMPLE 2 Converting to scientific notation

Convert the following to scientific notation.

a 0.00000452

b 12260000

Steps	Working
a 1 Count the number of times needed to move the decimal point to attain a number between 1 and 10. Has it moved left or right? **2** Write the answer.	
b 1 Count the number of times needed to move the decimal point to attain a number between 1 and 10. Has it moved left or right? **2** Write the answer.	

MATCHED EXAMPLE 3 — Converting from scientific notation

Convert the following from scientific notation.

a 2.35×10^{-3}

b 6.2435×10^{5}

Steps	Working
a 1 What is the power of 10? Does the decimal point need to move left or right? Do zeros need to be inserted? Or do the multiplication using CAS, if necessary. 2 Write the answer.	
b 1 What is the power of 10? Does the decimal point need to move left or right? Do zeros need to be inserted? Or do the multiplication using CAS, if necessary. 2 Write the answer.	

MATCHED EXAMPLE 4 Using Pythagoras' theorem to find unknown sides

For each of the following right-angled triangles, find the unknown values
  i correct to two decimal places
  ii correct to two significant figures.

a

(triangle with legs 10 m and 12 m, unknown x)

b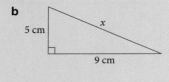

c

Steps	Working
a **i** **1** Identify the hypotenuse, c. The other two sides are a and b. Use Pythagoras' theorem to find the unknown side, using CAS if necessary.	
2 Write your answer in the required units and round to the required level of accuracy.	
ii Write your answer in the required units and round to the required level of accuracy.	
b **i** **1** Identify the hypotenuse, c. The other two sides are a and b. Use Pythagoras' theorem to find the unknown side, using CAS if necessary.	
2 Write your answer in the required units and round to the required level of accuracy.	
ii Write your answer in the required units and round to the required level of accuracy.	
c **i** **1** Identify the hypotenuse, c. The other two sides are a and b. Use Pythagoras' theorem to find the unknown side, using CAS if necessary.	

TI-Nspire **ClassPad**

  2 Write your answer in the required units and round to the required level of accuracy.

ii Write your answer in the required units and round to the required level of accuracy.

MATCHED EXAMPLE 5 — Using Pythagoras' theorem with shapes that contain right-angled triangles

Calculate the value of x, correct to one decimal place.

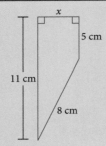

Steps | **Working**

1. Identify the right-angled triangle.
   Use the above diagram.

2. Label the sides of the triangle using the information given.
   Use the above diagram.

3. Identify the hypotenuse, c. The other two sides are a and b. Use Pythagoras' theorem to find the unknown side, using CAS if necessary.

4. Write your answer in the required units and round to the required level of accuracy.

MATCHED EXAMPLE 6 — Solving problems using Pythagoras' theorem in two dimensions

A sandwich with one side length of 7.62 cm is cut in half across the diagonal of length 13.74 cm. What is the length of the other side, to the nearest millimetre?

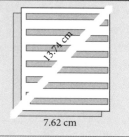

7.62 cm

Steps	Working
1 Identify the right-angled triangle and label the unknown side x. Label the above diagram.	
2 Use Pythagoras' theorem to find the unknown side, using CAS if necessary.	
3 Convert to the required units, round to the required accuracy and write the answer.	

MATCHED EXAMPLE 7 | Solving problems using Pythagoras' theorem in three dimensions

For this cycle ramp, what is the distance (in centimetres) a cyclist would travel if they cycled directly up the centre of the ramp to the nearest centimetre?

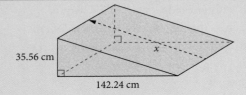

Steps	Working
1 Picture the diagram in three dimensions and find the relevant right-angled triangle with one unknown side. Label the diagram in the question.	
2 Redraw the right-angled triangle separately, labelling the two known sides and one unknown side.	
3 Use Pythagoras' theorem to find the unknown side, using CAS if necessary.	
4 Write your answer in the required units and round to the required level of accuracy.	

MATCHED EXAMPLE 8 — Solving two-step problems using Pythagoras' theorem in three dimensions

This rectangular box has a square base. The lengths of AH and CH are 12 cm and 9 cm, respectively.

a Find the length of AC correct to three significant figures.

b Use your answer to part **a** to find x, the length of AD, correct to three significant figures.

Steps	Working
a 1 Picture the diagram in three dimensions and find the relevant right-angled triangle with one unknown side. Use the diagram in the question.	
2 Redraw the right-angled triangle separately, labelling the two known sides and one unknown side.	
3 Use Pythagoras' theorem to find the unknown side, using CAS if necessary.	
4 Write your answer in the required units and round to the required level of accuracy.	
b 1 Picture the diagram in three dimensions and find the relevant right-angled triangle, which includes the side whose length we found in part **a**. Use the diagram in the question.	
2 Redraw the right-angled triangle separately, labelling the sides. Include the unrounded answer from part **a**.	
3 Use Pythagoras' theorem and the unrounded answer to part **a** to find x, using CAS if necessary.	
4 Write your answer in the required units and round to the required level of accuracy.	

MATCHED EXAMPLE 9 — Calculating the perimeter and area of quadrilaterals, triangles and circles

For each of the following shapes, calculate
  i the perimeter to the nearest centimetre
  ii the area to the nearest square centimetre.

a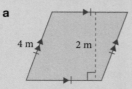

b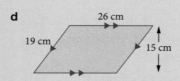

c
(circle with 15 cm shown as diameter)

d
(parallelogram with 26 cm, 19 cm, 15 cm)

Steps — **Working**

a i 1 Identify the shape.
  2 State the values from the diagram.
     Use the perimeter formula for that shape.

  3 Write your answer in the required units and round to the required level of accuracy.

 ii 1 Use the area formula for that shape.

  2 Write your answer in the required units and round to the required level of accuracy.

b i 1 Identify the shape.
  2 State the values from the diagram.
     Use the perimeter formula for that shape.

  3 Write your answer in the required units and round to the required level of accuracy.

 ii 1 Use the area formula for that shape.

  2 Write your answer in the required units and round to the required level of accuracy.

c **i** **1** Identify the shape.
  2 State the values from the diagram.
     Use the perimeter formula for that shape.

  3 Write your answer in the required units and round to the required level of accuracy.

ii **1** Use the area formula for that shape.

  2 Write your answer in the required units and round to the required level of accuracy.

d **i** **1** Identify the shape.
  2 State the values from the diagram.
     Use the perimeter formula for that shape.

  3 Write your answer in the required units and round to the required level of accuracy.

ii **1** Use the area formula for that shape.

  2 Write your answer in the required units and round to the required level of accuracy.

MATCHED EXAMPLE 10 | Calculating the perimeter and area of sectors

Calculate the following correct to three significant figures for each of the sectors shown.

　i arc length　　　　ii perimeter　　　　iii area

a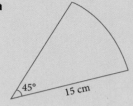

b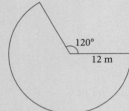

Steps | **Working**

a i 1 State the values from the diagram.
　　　　Use the arc length formula.

　　　2 Write your answer in the required units and round to the required level of accuracy.

　ii 1 State the values from the diagram.
　　　　Use the sector perimeter formula, using unrounded values where necessary.

　　　2 Write your answer in the required units and round to the required level of accuracy.

　iii 1 State the values from the diagram.
　　　　Use the sector area formula, using unrounded values where necessary.

　　　2 Write your answer in the required units and round to the required level of accuracy.

b i 1 State the values from the diagram.
　　　　Use the arc length formula.

　　　2 Write your answer in the required units and round to the required level of accuracy.

　ii 1 State the values from the diagram.
　　　　Use the sector perimeter formula, using unrounded values where necessary.

　　　2 Write your answer in the required units and round to the required level of accuracy.

▶ iii 1 State the values from the diagram.
Use the sector area formula, using unrounded values where necessary.

2 Write your answer in the required units and round to the required level of accuracy.

MATCHED EXAMPLE 11 — Calculating the perimeter and area of composite shapes

For each of the following shapes, calculate

i the perimeter   ii the area.

Give your answers using the units stated in the diagram to two significant figures in each case.

a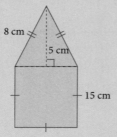

b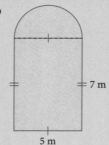

Steps   **Working**

a i 1 Identify the shapes that make up the composite shape.

    2 Calculate the missing lengths.
       Use the diagram in the question.

    3 Add all the lengths ignoring the lengths of the dashed lines.

    4 Write your answer in the required units and round to the required level of accuracy.

 ii 1 Separate the shapes that make up the composite shape.

    2 Use the area formulas for those shapes.

    3 Add the areas. Write your answer in the required units and round to the required level of accuracy.

b i 1 Identify the shapes that make up the composite shape.

   2 Calculate the missing lengths.
   Use the diagram in the question.

   3 Add all the lengths ignoring the length of the dashed line.

   4 Write your answer in the required units and round to the required level of accuracy.

ii 1 Separate the shapes that make up the composite shape.

   2 Use the area formulas for those shapes.

   3 Add the areas. Write your answer in the required units and round to the required level of accuracy.

MATCHED EXAMPLE 12 — Calculating the area of composite shapes where a shape is removed

Find the following shaded areas. Give your answers using the units stated in the diagram to three significant figures in each case.

a

16 cm, 10 cm, 3 m (circle)

b

6.2 m, 2.2 m (diamond)

Steps	Working
a i 1 Separate the shapes that make up the composite shape.	
2 Use the area formulas for those shapes.	
3 Subtract the empty space area from the other area. Write your answer in the required units and round to the required level of accuracy.	
b i 1 Separate the shapes that make up the composite shape.	
2 Use the area formulas for those shapes.	
3 Subtract the empty space area from the other area. Write your answer in the required units and round to the required level of accuracy.	

144 Nelson VICmaths General Mathematics 11 Mastery Workbook

MATCHED EXAMPLE 13 — Applying perimeter and area formulas

A square-shaped garden of length 25 m has a circle-shaped land of radius 10 m in its middle to plant a bed of roses.

a A fence is to be built all the way around the garden with two gates, each 1 m wide, and around the circle-shaped land for the bed of roses. The gates cost $110 each and the fencing costs $38 per metre. Calculate the total cost of fencing the garden and the rose bed, correct to the nearest dollar.

b If mulch to cover the garden costs $2 per square metre, calculate the cost to cover the garden with mulch, correct to the nearest dollar.

Steps	Working
a 1 Sketch the shape, showing the measurements. Decide if it is a perimeter or area problem.	
2 Add all the lengths. Add or delete any other given amounts. Write your unrounded answer in the required units.	
3 Calculate the cost and write the answer to the required level of accuracy.	
b 1 Decide if it is a perimeter or area problem. Separate the shapes that make up the composite shape.	
2 Use the area formulas for those shapes.	
3 Add all the areas. Write your unrounded answer in the required units.	
4 Calculate the cost and write the answer to the required level of accuracy.	

MATCHED EXAMPLE 14 Calculating the volume and capacity of prisms and cylinders

Calculate the volume (V) and capacity (C) of each of the following, rounding your answer to three significant figures.

a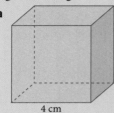

b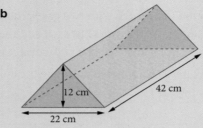

c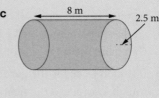

Steps	Working
a 1 Identify the object. Use a formula, or volume = base area × length if it is a prism we have no formula for.	
2 Write the volume, including units, and round to the required level of accuracy.	
3 Convert volume to capacity, including units, and round to the required level of accuracy.	
b 1 Identify the object. Use a formula, or volume = base area × length if it is a prism we have no formula for.	
2 Write the volume, including units, and round to the required level of accuracy.	
3 Convert volume to capacity, including units, and round to the required level of accuracy.	
c 1 Identify the object. Use a formula, or volume = base area × length if it is a prism we have no formula for.	
2 Write the volume, including units, and round to the required level of accuracy.	
3 Convert volume to capacity, including units, and round to the required level of accuracy.	

MATCHED EXAMPLE 15 — Calculating the volume and capacity of pyramids, cones and spheres

Calculate the volume (V) and capacity (C) of each of the following, rounding your answer to two significant figures.

a 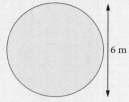   b 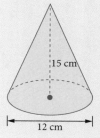   c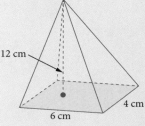

Steps	Working
a 1 Identify the object. Use a formula, or volume = $\frac{1}{3}$ × area of base × height if it is a pyramid we have no formula for.	
2 Write the volume, including units, and round to the required level of accuracy.	
3 Convert volume to capacity, including units, and round to the required level of accuracy.	
b 1 Identify the object. Use a formula, or volume = $\frac{1}{3}$ × area of base × height if it is a pyramid we have no formula for.	
2 Write the volume, including units, and round to the required level of accuracy.	
3 Convert volume to capacity, including units, and round to the required level of accuracy.	
c 1 Identify the object. Use a formula, or volume = $\frac{1}{3}$ × area of base × height if it is a pyramid we have no formula for.	
2 Write the volume, including units, and round to the required level of accuracy.	
3 Convert volume to capacity, including units, and round to the required level of accuracy.	

MATCHED EXAMPLE 16 | Calculating the volume of composite objects

Calculate the volume of the following, giving your answers to the nearest whole unit.

a 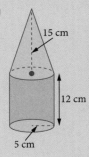   b 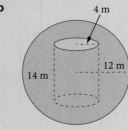   c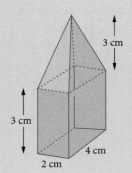

Steps	Working
a 1 Identify the objects that make up the composite object. Use the formulas needed, calculating any missing values.	
2 Calculate the total volume by adding or subtracting the volumes. Write your answer in the required units and round to the required level of accuracy.	
b 1 Identify the objects that make up the composite object. Use the formulas needed, calculating any missing values.	
2 Calculate the total volume by adding or subtracting the volumes. Write your answer in the required units and round to the required level of accuracy.	

c 1 Identify the objects that make up the composite object.

Use the formulas needed, calculating any missing values.

2 Calculate the total volume by adding or subtracting the volumes. Write your answer in the required units and round to the required level of accuracy.

MATCHED EXAMPLE 17 | Using surface area formulas

Calculate the surface area (SA) of each of the following, rounding your answer to two significant figures.

a

b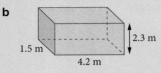

Steps	Working
a 1 Identify the object and the shapes that make up the net. Use the formula needed, calculating any missing values.	
2 Write your answer in the required units and round to the required level of accuracy.	
b 1 Identify the object and the shapes that make up the net. Use the formula needed, calculating any missing values.	
2 Write your answer in the required units and round to the required level of accuracy.	

MATCHED EXAMPLE 18 — Applying surface area formulas

Calculate the surface area (SA) of each of the following, rounding your answer to two significant figures.

a A rectangular shoe box without a lid

b A pyramid-shaped lampshade

c An ice cream cone

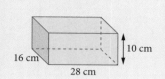

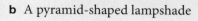

Steps	Working
a 1 Identify the object and the shapes that make up the net, drawing the net if necessary. Adapt the longer version of the formulas needed, calculating any missing values.	
2 Write your answer in the required units and round to the required level of accuracy.	
b 1 Identify the object and the shapes that make up the net, drawing the net if necessary. Adapt the longer version of the formulas needed, calculating any missing values.	
2 Write your answer in the required units and round to the required level of accuracy.	
c 1 Identify the object and the shapes that make up the net, drawing the net if necessary. Adapt the longer version of the formulas needed, calculating any missing values.	
2 Write your answer in the required units and round to the required level of accuracy.	

MATCHED EXAMPLE 19 — Working with scale factors

For each of these pairs of similar shapes, find

　i　the scale factor k
　ii　the value of x, rounded to one decimal place.

a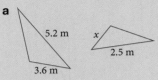

b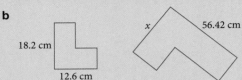

Steps	Working
a i Find a pair of matching lengths and use $$k = \frac{\text{any length of second shape}}{\text{matching length of first shape}}$$ ii Use the scale factor to find and solve an equation for the unknown, using CAS if necessary. Write your answer to the required accuracy.	
b i Find a pair of matching lengths and use $$\text{Scale factor} = \frac{\text{any length of second shape}}{\text{matching length of first shape}}$$ ii Use the scale factor to find and solve an equation for the unknown, using CAS if necessary. Write your answer to the required accuracy.	

MATCHED EXAMPLE 20 | Identifying similar shapes

State whether the following pairs of shapes are similar, giving a reason.

a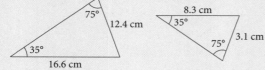

b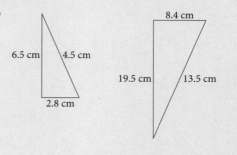

c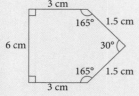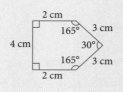

Steps	Working
a 1 Identify the shape. 2 Calculate the scale factors of the matching sides in order, starting with the largest values in each shape. If necessary, look at the angles.	
b 1 Identify the shape. 2 Calculate the scale factors of the matching sides in order, starting with the largest values in each shape. If necessary, look at the angles.	
c 1 Identify the shape. 2 Calculate the scale factors of the matching sides in order, starting with the largest values in each shape. If necessary, consider the angles.	

MATCHED EXAMPLE 21 — Scaling areas and volumes

A cone with radius of 12 cm is enlarged to produce a similar cone with radius 48 cm.

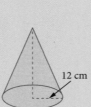
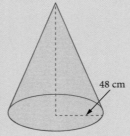

Find

a the scale factor k

b the surface area of the larger cone if the surface area of the smaller cone is 124 cm^2

c the volume of the smaller cone if the volume of the larger cone is 608 cm^3.

Round your answers to three significant figures.

Steps	Working
a Use $k = \dfrac{\text{any length of second object}}{\text{matching length of first object}}$	
b 1 Use surface area of second object = $k^2 \times$ surface area of first object 2 Write your answer in the required units and round to the required level of accuracy.	
c 1 Use volume of second object = $k^3 \times$ volume of first object 2 Write your answer in the required units and round to the required level of accuracy.	

APPPLICATIONS OF TRIGONOMETRY

CHAPTER 10

MATCHED EXAMPLE 1 — Finding an unknown side of a right-angled triangle

Find the value of the pronumeral, correct to two decimal places, in each right-angled triangle.

a
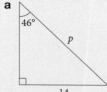

b

SB
p. 453

Steps | **Working**

a 1 Indicate the sides of the right-angled triangle with the letters O, A and H to show opposite, adjacent and hypotenuse.

Use the diagram in the question.

2 Identify whether the labelled sides are opposite, adjacent or hypotenuse and select the matching trigonometric ratio.

3 Substitute the known values and solve the equation for the unknown to the required level of accuracy, using the **solve** function on CAS if necessary.

TI-Nspire

ClassPad

b 1 Indicate the sides of the right-angled triangle with the letters O, A and H to show opposite, adjacent and hypotenuse.

Use the diagram in the question.

2 Identify whether the labelled sides are opposite, adjacent or hypotenuse and select the matching trigonometric ratio.

3 Substitute the known values and solve the equation for the unknown to the required level of accuracy, using the **solve** function on CAS if necessary.

TI-Nspire   **ClassPad**

MATCHED EXAMPLE 2 — Solving problems involving an unknown side of a right-angled triangle

A ladder is inclined at 25° to the wall. If the distance of the foot of the ladder from the bottom of the wall is 3 m, what is the height of the ladder? Round your answer correct to one decimal place.

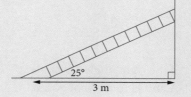

Steps	Working
1 Redraw the triangle using the given values. Represent the length of the ladder by x.	
2 Indicate the sides of the right-angled triangle with the letters O, A and H to show opposite, adjacent and hypotenuse. Use the above diagram.	
3 Identify whether the labelled sides are opposite, adjacent or hypotenuse and select the matching trigonometric ratio.	
4 Substitute the known values and solve the equation for the unknown, using the **solve** function on CAS if necessary. **TI-Nspire** **ClassPad**	
5 Write your answer in the required units and round to the required level of accuracy.	

MATCHED EXAMPLE 3 Finding an unknown angle in a right-angled triangle

Find the value of θ, correct to the nearest degree.

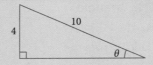

Steps	Working
1 Indicate the sides of the right-angled triangle with the letters O, A and H to show opposite, adjacent and hypotenuse. Use the diagram in the question.	
2 Identify whether the labelled sides are opposite, adjacent or hypotenuse and select the matching trigonometric ratio.	
3 Substitute the known values and solve for θ with CAS, using the inverse trigonometric function and rounding to the nearest degree.	

TI-Nspire

ClassPad

Press **trig** to open the mini-palette to access the inverse trigonometry functions.

MATCHED EXAMPLE 4 — Solving problems involving an unknown angle of a right-angled triangle

A skateboard kicker ramp has a length of 3.5 m and covers a horizontal distance of 2.8 m. What is its angle of inclination, θ, correct to the nearest degree?

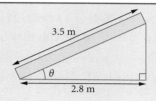

Steps	Working
1 Redraw the triangle using the given values. Represent the required angle by θ.	
2 Indicate the sides of the right-angled triangle with the letters O, A and H to show opposite, adjacent and hypotenuse. Use the above diagram.	
3 Identify whether the labelled sides are opposite, adjacent or hypotenuse and select the matching trigonometric ratio.	
4 Substitute the known values and solve for θ with CAS, using the inverse trigonometric function and rounding to the nearest degree.	
5 Write your answer in the required units and round to the required level of accuracy.	

MATCHED EXAMPLE 5 — Solving problems involving angles of elevation and depression

Find each of the following, to the nearest whole unit, by drawing a diagram.

a Meena stands 100 m from the base of a 48 m tall chimney and looks up at the top of the chimney. Find the angle of elevation θ of the top of the chimney from Meena.

b Ann is standing on a 4.5 m tall building looking down at a cafe on the opposite side of the building. The angle of depression of the cafe is 32°. Find the distance d from the cafe to the base of the building.

Steps	Working
a 1 Draw a diagram with all the measurements including the angle. If necessary, use angle of elevation X to Y = angle of depression Y to X. Indicate the sides of the right-angled triangle with the letters O, A and H to show opposite, adjacent and hypotenuse. **2** Identify whether the labelled sides are opposite, adjacent or hypotenuse and select the matching trigonometric ratio. **3** Substitute the known values and solve with CAS, rounding to the required level of accuracy.	
b 1 Draw a diagram with all the measurements including the angle. If necessary, use angle of elevation X to Y = angle of depression Y to X. Indicate the sides of the right-angled triangle with the letters O, A and H to show opposite, adjacent and hypotenuse. **2** Identify whether the labelled sides are opposite, adjacent or hypotenuse and select the matching trigonometric ratio. **3** Substitute the known values and solve with CAS. Write your answer in the required units and round to the required level of accuracy.	

MATCHED EXAMPLE 6 — Calculating three-figure bearings

For each of the following, calculate the three-figure bearing of the point A and draw a diagram showing the angle.

a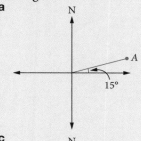

b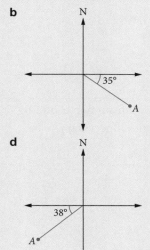

c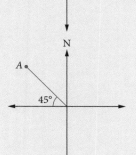

d

Steps	Working
a 1 Find the angle clockwise from north, adding or subtracting 90° angles if necessary.	
2 Write your answer, making sure the answer has three figures. Draw a diagram showing the angle.	
b 1 Find the angle clockwise from north, adding or subtracting 90° angles if necessary.	

2 Write your answer, making sure the answer has three figures.

c 1 Find the angle clockwise from north, adding or subtracting 90° angles if necessary.

2 Write your answer, making sure the answer has three figures.

d 1 Find the angle clockwise from north, adding or subtracting 90° angles if necessary.

2 Write your answer, making sure the answer has three figures.

MATCHED EXAMPLE 7 Applying three-figure bearings

A barrage balloon travelled 2.8 km due west and then 4.2 km north.

a What is its three-figure bearing from its starting point, to the nearest degree?

b How far is it from its starting point? Round your answer to one decimal place.

Steps	Working
a 1 Draw a diagram showing the information as a right-angled triangle. Include θ as an unknown angle in the triangle.	
2 Use the triangle to find θ. Identify whether the labelled sides are opposite, adjacent or hypotenuse and select the matching trigonometric ratio.	
3 Substitute the known values and solve with CAS, rounding to the nearest whole unit.	
4 Mark the three-figure bearing angle on the diagram. Find the angle clockwise from north, adding or subtracting 90° angles if necessary. Write the answer to the required level of accuracy.	
b 1 Use Pythagoras' theorem to find d, the distance from the starting point.	
2 Write your answer in the required units and round to the required level of accuracy.	

MATCHED EXAMPLE 8 Using the sine rule for non-right-angled triangles

Find the unknown value for each of the following to the nearest whole unit.

a

b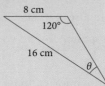

Steps	Working
a 1 Label the angles A, B, C and label the sides opposite the angles a, b, c. Use the diagram in the question.	
2 List the known values and the value that needs to be found. Select the equality needed from the sine rule: $$\frac{a}{\sin(A)} = \frac{b}{\sin(B)} = \frac{c}{\sin(C)}$$	
3 Substitute the values into the sine rule and solve using CAS. **TI-Nspire**	**ClassPad**
4 Write your answer in the required units and round to the required level of accuracy.	
b 1 Label the angles A, B, C and label the sides opposite the angles a, b, c. Use the diagram in the question.	
2 List the known values and the value that needs to be found. Select the equality needed from the sine rule: $$\frac{a}{\sin(A)} = \frac{b}{\sin(B)} = \frac{c}{\sin(C)}$$	
3 Substitute the values into the sine rule and solve using CAS.	

TI-Nspire **ClassPad**

4 Choose the correct answer by deciding from the diagram if θ is less than 90° or greater than 90°.

Write your answer in the required units and round to the required level of accuracy.

MATCHED EXAMPLE 9 — Solving problems using the sine rule

Ron observes a tree at a 32° angle of elevation. Walking 12 m towards the tree, he finds that the angle of elevation increases to 55°.

a Draw the two triangles involved and show the sizes of all the angles.

b Hence evaluate the height, h metres, of the tree correct to two decimal places.

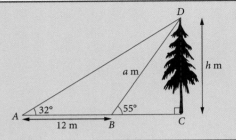

Steps	Working
a 1 Draw the two triangles separately showing all the information given in the diagram.	
2 Use the fact that a straight angle is 180° and the angles of a triangle sum to 180°. Use the above diagram.	
b 1 Select the equality needed from the sine rule to find the unknown in the first triangle: $$\frac{a}{\sin(A)} = \frac{b}{\sin(B)} = \frac{c}{\sin(C)}$$ Solve using CAS.	

> 2 Use the unknown length found and the sine rule to calculate the required value from the second triangle.
>
>   Solve using CAS.
>
> 3 Write your answer in the required units and round to the required level of accuracy.

MATCHED EXAMPLE 10 Using the cosine rule for non-right-angled triangles

Find the unknown value for each of the following to the nearest whole unit.

a

b

Steps	**Working**
a 1 Label the shown angle A, then label the other angles B and C and the sides opposite the angles a, b, c. Use the diagram in the question. **2** List the known values and the value that needs to be found. Select the required version of the cosine rule. $a^2 = b^2 + c^2 - 2bc\cos(A)$ or $\cos(A) = \dfrac{b^2 + c^2 - a^2}{2bc}$ Solve using CAS. **3** Write your answer in the required units and round to the required level of accuracy.	
b 1 Label the shown angle A, then label the other angles B and C and the sides opposite the angles a, b, c. Use the diagram in the question. **2** List the known values and the value that needs to be found. Select the required version of the cosine rule. $a^2 = b^2 + c^2 - 2bc\cos(A)$ or $\cos(A) = \dfrac{b^2 + c^2 - a^2}{2bc}$ Solve using CAS. **3** Write your answer in the required units and round to the required level of accuracy.	

MATCHED EXAMPLE 11 Solving problems involving non-right-angled triangles

Rachel is going on a three-day hike. She starts at campsite X and walks for 12 km on a bearing of 145° to camp site Y. On the second day, she walks 18 km due west to campsite Z. On the third day, Rachel plans to return to campsite X.

a Find the distance, to the nearest kilometre, that Rachel needs to travel to return to campsite X.

b What three-figure bearing does Rachel need to travel on in order to return to campsite X? Express your answer correct to the nearest degree.

Steps	Working
a 1 Draw a diagram showing the information as a triangle, including the unknown we are asked to find.	
2 Calculate an angle in the triangle.	
3 Is it a right-angled triangle? Is a side and its opposite angle known?	
4 Redraw the triangle. Label the shown angle A, then label the other angles B and C, and the sides opposite the angles a, b, c.	
5 List the known values and the value that needs to be found. Select the required version of the cosine rule. $a^2 = b^2 + c^2 - 2bc \cos(A)$ or $\cos(A) = \dfrac{b^2 + c^2 - a^2}{2bc}$ Solve using CAS.	

▶		**6** Write your answer in the required units and round to the required level of accuracy.
	b 1	Add north to the bearing starting point on the redrawn triangle, include the unrounded distance found in part a, and show the angle to find.
	2	Is it a right-angled triangle?
		Is a side and its opposite angle known?
	3	Select the equality needed from the sine rule to find the unknown in the first triangle:
		$$\frac{a}{\sin(A)} = \frac{b}{\sin(B)} = \frac{c}{\sin(C)}$$
		Solve using CAS, rounding to the nearest degree.
	4	Answer the question by calculating the three-figure bearing.

Answers

Worked solutions available on Nelson MindTap.

CHAPTER 1

MATCHED EXAMPLE 1

a Numerical, discrete and ratio
b Numerical, continuous and ratio
c Categorical and nominal
d Categorical and nominal
e Categorical and ordinal
f Numerical, continuous and ratio

MATCHED EXAMPLE 2

a i 2 m   ii 2 m   iii 5 m
b i 23 cakes   ii 23 cakes   iii 7 cakes

MATCHED EXAMPLE 3

Lolly type	Frequency	Percentage
Cheekies	5	20%
Sour Ears	7	28%
Fads	7	28%
Minties	3	12%
Jaffas	3	12%
Total	25	100%

MATCHED EXAMPLE 4

80–<100

MATCHED EXAMPLE 5

a Day 2   b 40 hours   c 10%

MATCHED EXAMPLE 6

a 8 intervals
b 27 adults
c 10.9%
d The histogram is approximately symmetric without a possible outlier.
e 164–<170 cm.

f
Height (kg)	Frequency
140–<146	2
146–<152	10
152–<158	14
158–<164	18
164–<170	20
170–<176	19
176–<182	15
182–<188	12
Total	110

MATCHED EXAMPLE 7

a min = 36, Q_1 = 51.5, median = 54.5, Q_3 = 62, max = 79

b 36, 49, 50, 51, | 52, 53, 53, 54, | 55, 59, 60, 60, | 64, 64, 78, 79

Lower quartile = 51.5   Median = 54.5   Upper quartile = 62

MATCHED EXAMPLE 8

74 is less than 82.5, so it is a possible outlier.
98 is less than 102.5, so it is *not* an outlier.
130 is greater than 102.5, so it is a possible outlier.

MATCHED EXAMPLE 9

a min = 120, Q_1 = 180, median = 270, Q_3 = 440, max = 620
b 50%   c 25%
d 25%   e 9
f −210   g 830

MATCHED EXAMPLE 10

a i   mode = 5 books
   ii range = 5 books
   iii median = 5
   iv $Q_1 = 4$
   v $Q_3 = 6.5$
   vi IQR = 2.5
b The distribution is approximately symmetric.

MATCHED EXAMPLE 11

a i   mode = 45 and 58
   ii range = 65
   iii median = 58
   iv $Q_1 = 48$
   v $Q_3 = 70$
   vi IQR = 22
b The 25 minutes may be an outlier.
   $Q_1 - 1.5 \times IQR = 48 - 1.5 \times 22 = 15$.
   Since 25 isn't less than 15, it is not an outlier.

MATCHED EXAMPLE 12

a
Camera 1		Camera 2
9 8 8 8 7 5 5 4	6	
9 9 8 8 7 0	7	
8 7 7 7 5	8	
6 5 0	9	4 5 6
	10	5 8
	11	4 4 5 6 8
	12	0 2 3 4 5 6 6
	13	0 3 6 7 7
0|9 = 90 km/h		13|6 = 136 km/h

b The data for Camera 2 is negatively skewed.

c Camera 1: median = 78, range = 32, IQR = 19
Camera 2: median = 121, range = 43, IQR = 12

d The second road has the greater speeding problem. The median speed for the first road is 78 km/h, which is 12 km/h *under* the 90 km/h speed limit. The median speed for the second road is 121 km/h, which is 11 km/h *above* the 110 km/h speed limit.

MATCHED EXAMPLE 13

a

Mia		Chaya
1 0	1	
6 5	1	
0 0 0	2	3 3
8 6 6	2	5 7 8
1 1 1 0	3	2 3 4 4
	3	5 6 7 7 8
6|2 = 26 flyers		2|3 = 23 flyers

b Chaya's data is negatively skewed.

c Chaya: median = 33.5, range = 15, IQR = 9
Mia: median = 23, range = 21, IQR = 14

d Chaya's median (33.5) is higher than Mia's median (23). Mia's range (21) and IQR (14) are considerably higher than Chaya's range (15) and IQR (9). This means Chaya's deliveries have less variability and are higher than Mia's, which indicates that Chaya is the better delivery person.

MATCHED EXAMPLE 14

a Forbes   **b** Laverton
c Warwick   **d** Warwick

e Warwick has noticeably higher average October temperatures than Laverton and Forbes. Warwick's median (12°C) is higher than Laverton's (9°C) and Forbes' (8°C).

MATCHED EXAMPLE 15

a **i** 12   **ii** 3.5
b **i** 16   **ii** 2.2

MATCHED EXAMPLE 16

a **i** $\bar{x} = 55.22$, $s = 2.34$
   ii 10
b **i** $\bar{x} = 47.40$, $s = 5.37$
   ii 9

MATCHED EXAMPLE 17

a 47.5%   **b** 2.5%
c 13.5%   **d** 0.15%

CHAPTER 2

MATCHED EXAMPLE 1

a **i** −1, 2, 5, 8, 11
   ii Add 3 to generate each new value.

b **i** 2, 6, 10, 14, 18
   ii Add 4 to generate each new value.

c **i** 1, 3, 9, 27, 81
   ii Multiply by 3 to generate each new value.

d **i** 1, −2, 4, −8, 16
   ii Multiply by −2 to generate each new value.

MATCHED EXAMPLE 2

a Oscillating
b Increasing
c Decreasing, limiting value
d Constant
e Decreasing

MATCHED EXAMPLE 3

a **i** 3
   ii $u_2 = 13$, $u_5 = 22$
   iii

n	0	1	2	3	4	5	6
u_n	7	10	13	16	19	22	25

   iv

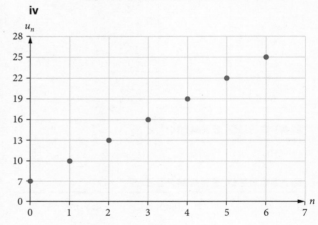

   v 3

The slope is the amount being added to generate each new value.

b **i** 2
   ii $u_2 = 0$, $u_5 = -6$
   iii

n	0	1	2	3	4	5	6
u_n	4	2	0	−2	−4	−6	−8

   iv

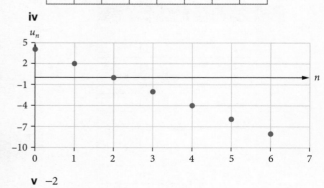

   v −2

The slope is the amount being added to generate each new value.

MATCHED EXAMPLE 4

a $a = 2$
$d = 3$

b 2, 5, 8, 11

MATCHED EXAMPLE 5

a **i** We are adding 4 to generate each new value, so this is an arithmetic sequence.
ii $u_n = 1 + 4n$
iii $u_{20} = 81$
$u_{100} = 401$

b **i** We are subtracting 3 to generate each new value, so this is an arithmetic sequence.
ii $u_n = 60 - 3n$
iii $u_{20} = 0$
$u_{100} = -240$

MATCHED EXAMPLE 6

a The fixed amount of interest paid for each year is $120.

b Mia's bank account balance after three years is $2360.
Mia's balance is first greater than $2400 after four years.
Total amount of interest earned after six years is $720.

c $u_0 = 2000$, $u_{n+1} = u_n + 120$

d

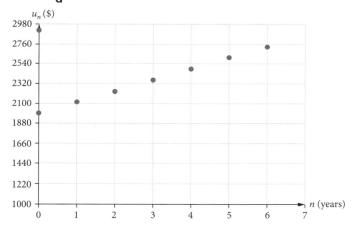

The points are in an increasing straight line.

MATCHED EXAMPLE 7

a $6200   **b** $3800

MATCHED EXAMPLE 8

a $180

b $u_0 = 3000$, $u_n = 3000 + 180n$

c $4800

MATCHED EXAMPLE 9

a $4000

b The value of the car after three years is $8000.

The value of the car first falls below $10 000 after three years.

The car depreciates to zero after five years.

c $u_0 = 20\,000$, $u_{n+1} = u_n - 4000$

d

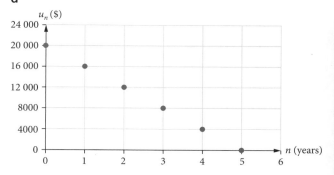

MATCHED EXAMPLE 10

a The refrigerator decreases by $1050 each year.

b $u_1 = 9450$
$u_2 = 8400$
$u_3 = 7350$

c 10%

MATCHED EXAMPLE 11

a $61 280

b $u_n = 383\,000 - 61\,280n$

c $137 880

d 7 years

MATCHED EXAMPLE 12

a The amount of depreciation is determined by applying a rate per unit of use: $50 every time the sewing machine is used to produce a garment.

b The value of the sewing machine after it produces four garments is $4800.
It will take five garments for the value of the sewing machine to first fall below $4800.

c $u_0 = 5000$, $u_{n+1} = u_n - 50$

d

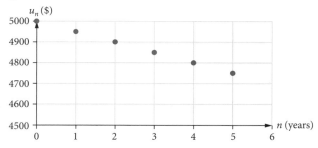

MATCHED EXAMPLE 13

a $u_n = 2000 - 0.50n$ **b** $1950

CHAPTER 3

MATCHED EXAMPLE 1

a **i** We are multiplying by 2 to generate each new value, so this is a geometric sequence.

  ii $R = \frac{110}{55} = 2, R = \frac{220}{110} = 2, R = \frac{440}{220} = 2$

  iii $u_2 = 220, u_4 = 880$

  iv

n	0	1	2	3	4	5
u_n	55	110	220	440	880	1760

  v

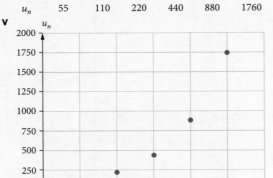

b **i** We are dividing by 3 to generate each new value $\left(\text{or multiplying by } \frac{1}{3}\right)$, so this is a geometric sequence.

  ii $R = \frac{18}{54} = \frac{1}{3}, R = \frac{6}{18} = \frac{1}{3}, R = \frac{2}{6} = \frac{1}{3}$

  iii $u_2 = 6, u_4 = \frac{2}{3}$

  iv

n	0	1	2	3	4	5
u_n	54	18	6	2	$\frac{2}{3}$	$\frac{2}{9}$

  v

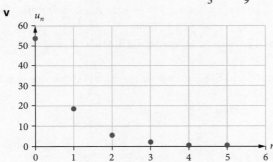

MATCHED EXAMPLE 2

a $a = 3$
   $R = -4$

b $3, -12, 48, -192$

MATCHED EXAMPLE 3

a **i** $\frac{64}{16} = 4, \frac{256}{64} = 4, \frac{1024}{256} = 4$

   We are multiplying by 4 to generate each new value, so this is a geometric sequence.

  ii $u_n = 16 \times 4^n$

  iii $u_5 = 16\,384, u_8 = 1\,048\,576$

b **i** We are multiplying by 0.5 to generate each new value, so this is a geometric sequence.

  ii $u_n = -500 \times (0.5)^n$

  iii $u_5 \approx -15.63$
   $u_8 \approx -1.95$

MATCHED EXAMPLE 4

a

	Compound		Simple	
n	Interest ($)	Value of investment ($)	Interest ($)	Value of investment ($)
0	–	8000	–	8000
1	$\frac{15}{100} \times 8000$ $= 1200$	$8000 + 1200 =$ 9200	$\frac{15}{100} \times 8000$ $= 1200$	$8000 + 1200$ $= 9200$
2	$\frac{15}{100} \times 9200$ $= 1380$	$9200 + 1380 =$ $10\,580$	$\frac{15}{100} \times 8000$ $= 1200$	$9200 + 1200$ $= 10\,400$
3	$\frac{15}{100} \times 10\,580$ $= 1587$	$10\,580 + 1587$ $= 12\,167$	$\frac{15}{100} \times 8000$ $= 1200$	$10\,400 + 1200$ $= 11\,600$
4	$\frac{15}{100} \times 12\,167$ $= 1825.05$	$12\,167 + 1825.05$ $= 13\,992.05$	$\frac{15}{100} \times 8000$ $= 1200$	$11\,600 + 1200$ $= 12\,800$

b $13\,992.05

c The compound interest investment has $1192.05 more.

MATCHED EXAMPLE 5

a **i** 12   **ii** 120
   iii $\frac{1}{2}\%$ **iv** $75.00

b **i** 52   **ii** 520
   iii $\frac{5}{26}\%$ **iv** $300.00

MATCHED EXAMPLE 6

a $u_0 = 25\,000, u_{n+1} = 1.011 u_n$

b $u_0 = 30\,000, u_{n+1} = 1.001 u_n$

c $u_0 = 35\,000, u_{n+1} = 1.0002 u_n$

MATCHED EXAMPLE 7

a 1%   **b** $u_n = 1.01^n \times 45\,000$
c $54\,908.55   **d** $9908.55

MATCHED EXAMPLE 8

	Depreciation after n years ($)	Value after n years ($)
0	–	70\,000
1	$\frac{20}{100} \times 70\,000 = 14\,000$	$70\,000 - 14\,000 = 56\,000$
2	$\frac{20}{100} \times 56\,000 = 11\,200$	$56\,000 - 11\,200 = 44\,800$
3	$\frac{20}{100} \times 44\,800 = 8960$	$44\,800 - 8960 = 35\,840$
4	$\frac{20}{100} \times 35\,840 = 7168$	$35\,840 - 7168 = 28\,672$
5	$\frac{20}{100} \times 28\,672 = 5734$	$28\,672 - 5734 = 22\,938$

b The school bus is depreciated by $8960 in the third year.

c $u_0 = 70\,000$, $u_{n+1} = 0.8u_n$

MATCHED EXAMPLE 9

a The annual percentage rate is 20%.

b The truck first depreciates to under $45 000 after three years.

MATCHED EXAMPLE 10

a The TV is being depreciated by 10% of its *value* each year. If it was flat rate depreciation, it would be depreciated by 10% of its *initial value* each year.

b $u_n = 0.9^n \times 8000$

c $4252

MATCHED EXAMPLE 11

a i $14.25 ii −$9.50

b i $100.70 ii $83.60

MATCHED EXAMPLE 12

a 20% b 15%

MATCHED EXAMPLE 13

a $90.91 b $76.47

MATCHED EXAMPLE 14

a i $68 182 ii $6818

b i $154 ii $14

MATCHED EXAMPLE 15

a i One burger costs $12.40.
  ii 50 burgers cost $620.

b i One packet has a mass of 525 grams.
  ii 30 packets of wheat flour have a total mass of 15 750 grams.

MATCHED EXAMPLE 16

a The total number of residents in the apartment is 300.

b There are 30 children in the apartment.

MATCHED EXAMPLE 17

a The cost of one gram in each case is:
  $0.0214, $0.0194, $0.0195

b The 500 g for $9.70 is the best value because it is the cheapest per gram.

MATCHED EXAMPLE 18

a 5333.3% b 763.3%

MATCHED EXAMPLE 19

a $1.18 b $1.22

c $1.27 d 10.4%

e Each year's inflation compounds with the previous year's inflation.

MATCHED EXAMPLE 20

The $210 pocket watch is now worth $145.54.

MATCHED EXAMPLE 21

a 1.25%

b i $4618.25 ii $4231.73 iii $3840.38

c $5331.00

d $331.00

MATCHED EXAMPLE 22

a $935 b $1158 c $1180

d $1133 e $1148.64

f The lowest cost is to wait 3 months and pay cash for the item when on sale.

CHAPTER 4

MATCHED EXAMPLE 1

a When $x = -2$, $y = -11$
  When $x = -1$, $y = -9$
  When $x = 0$, $y = -7$
  When $x = 1$, $y = -5$
  When $x = 2$, $y = -3$

b
x	−2	−1	0	1	2
y	−11	−9	−7	−5	−3

c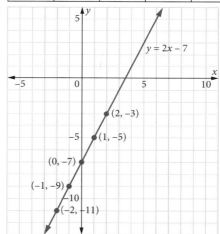

MATCHED EXAMPLE 2

a $(3, 14)$ doesn't lie on the line $y = -5x + 3$ since $14 \neq -5 + 3$.

b $(3, 14)$ lies on the line $y = 5x - 1$ since $14 = 15 - 1$.

MATCHED EXAMPLE 3

a i The slope is not defined.

b i The line is sloping down from left to right, so the slope is negative.
  ii Slope = −3

c i The slope is 0.

d i The line is sloping up from left to right, so the slope is positive.
  ii Slope = 1

MATCHED EXAMPLE 4
a slope = 3   b slope = −2   c slope = 3

MATCHED EXAMPLE 5
a $y = 3$.   b $y = -x + 3$ c $y = 2x + 3$

MATCHED EXAMPLE 6
a The vertical intercept is 5.
b slope = 0.25
c $C = 0.25m + 5$
d The entry cost is $5.
e The rate of charge per ride is $0.25.
f The total cost of going on 25 rides is $11.25.
g 40 rides have a total cost of $15.

MATCHED EXAMPLE 7
a $C = 32n + 820$
b The total cost if 7 students attended is $1044.
c The games would not have been planned if so few students were coming.
d The total cost if 900 students attended is $29 620.
e The venue has a maximum capacity of 150, so grade 11 students would have had to find another venue and the cost equation would have been different.

MATCHED EXAMPLE 8
a revenue = $11.25n$
b profit = $4.25n - 180$
c 150 vegetable pizzas would make a profit of $457.50.
d 40 vegetable pizzas would make a loss of $10.
e The profit needs to be *at least* $1500 and selling 395 vegetable pizzas won't quite make that profit. So, the number of vegetable pizzas that need to be sold is 396.

MATCHED EXAMPLE 9
a The coordinates of the x-intercept are $(3, 0)$.
b The coordinates of the y-intercept are $\left(0, -4\frac{1}{2}\right)$.
c
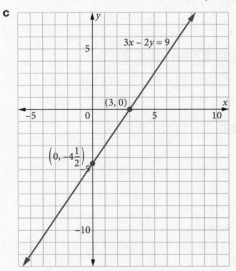

MATCHED EXAMPLE 10
Daphne purchased 7 pastries and 2 cakes.

MATCHED EXAMPLE 11
a The graph is made up of four different line segments and each line segment has a different slope.
b After 10 hours, the oil container holds 12.5 litres.
c The oil container filled to capacity after 25 hours.
d The container fills the fastest from approximately 1.5 hours to 4 hours after the start of the oil machine.
e The container is filling at a rate of 0.5 litres per hour in the last 15 hours.

MATCHED EXAMPLE 12
a Catherine started her journey 1 km from her own house.
b The slope of the line segment after 15 minutes is less than the slope of the line segment before 15 minutes.
c The horizontal line indicates that Catherine is the same distance (4 km) from her house during this time.
d Catherine stayed Amy's house for 45 minutes.
e The distance between Catherine's house and Amy's house is 4 km.
f Catherine's speed = 4 km/h

MATCHED EXAMPLE 13
a   i Riding for 40 minutes will cost $10.
   ii 5 hours of riding will cost $20.
   iii 7 hours of riding will cost $25.
   iv 9 hours 30 minutes of riding will cost $35.
b For $20 the person can ride for a time that is more than 3 hours up to 5 hours.
c After 8 hours, the riding charge remains constant at $35, so the maximum fare is $35.

CHAPTER 5

MATCHED EXAMPLE 1
a
$$M = \begin{bmatrix} 60 & 50 & 75 & 20 & 90 \\ 150 & 120 & 90 & 130 & 100 \\ 30 & 80 & 60 & 40 & 80 \\ 180 & 150 & 130 & 160 & 155 \end{bmatrix}$$

The order of M is 4×5.

M has 20 elements.

b $[90]$

The order is 1×1

c $\begin{bmatrix} 60 & 50 & 75 & 20 & 90 \end{bmatrix}$

d $\begin{bmatrix} 150 \\ 120 \\ 90 \\ 130 \\ 100 \end{bmatrix}$

e $\begin{bmatrix} 50 & 120 & 80 & 150 \end{bmatrix}$

f $\begin{bmatrix} 295 \\ 590 \\ 290 \\ 775 \end{bmatrix}$

g

	Cake	Coffee	Ice cream	Croissant
Monday	60	150	30	180
Tuesday	50	120	80	150
Wednesday	75	90	60	130
Thursday	20	130	40	160
Friday	90	100	80	155

MATCHED EXAMPLE 2

a 4×4, square matrix

b 1×4, row matrix

c 3×3, square matrix, identity matrix

d 5×5, square matrix, zero matrix

e 4×1, column matrix

MATCHED EXAMPLE 3

a $\begin{bmatrix} -1 & 0 \\ -6 & -2 \\ 0 & 1 \end{bmatrix}$

b $\begin{bmatrix} 4 & -3 & 4 & 2 \end{bmatrix}$

c $\begin{bmatrix} 6 & -4 & 8 & 2 \end{bmatrix}$

d $\begin{bmatrix} \frac{1}{2} & 1 \\ -\frac{1}{2} & -1 \\ \frac{1}{2} & 1 \end{bmatrix}$

e $\begin{bmatrix} -1 & -4 \\ 8 & 6 \\ -2 & -5 \end{bmatrix}$

f $2B + A$ isn't defined because the order of $2B$ is 4×1 and the order of A is 3×2. Matrices must have the same order for addition to be possible.

MATCHED EXAMPLE 4

a $x = 2, y = 1, z = -7$

b $x = 7, y = 14, z = 4$

MATCHED EXAMPLE 5

a $2.2 \times \begin{bmatrix} 6799 \\ 8899 \\ 9999 \\ 10\,980 \end{bmatrix}$

b $2.2 \times \begin{bmatrix} 6799 \\ 8899 \\ 9999 \\ 10\,980 \end{bmatrix} + \begin{bmatrix} 85 \\ 85 \\ 85 \\ 85 \end{bmatrix}$

c $2.2 \times \begin{bmatrix} 6799 \\ 8899 \\ 9999 \\ 10\,980 \end{bmatrix} + \begin{bmatrix} 85 \\ 85 \\ 85 \\ 85 \end{bmatrix} - \begin{bmatrix} 500 \\ 500 \\ 500 \\ 1000 \end{bmatrix}$

MATCHED EXAMPLE 6

a BC is not defined.

b $\begin{bmatrix} 7 & 11 \\ 10 & 8 \\ 6 & 3 \end{bmatrix}$

c $\begin{bmatrix} 6 & 2 & 2 \\ 15 & 5 & 5 \\ 24 & 8 & 8 \end{bmatrix}$

d $[19]$

e C^2 is not defined.

f $\begin{bmatrix} 11 & 7 \\ 7 & 11 \end{bmatrix}$

g CA is not defined.

MATCHED EXAMPLE 7

a **i** Yes. det$(A) = -10$

ii $\begin{bmatrix} -0.2 & 0.4 \\ 0.3 & -0.1 \end{bmatrix}$

b **i** Yes. det$(B) = 0$

ii B^{-1} doesn't exist.

c **i** No. det(C) doesn't exist.

ii Only square matrices have inverses, so C^{-1} doesn't exist.

d **i** Yes. det$(D) = 1$

ii $\begin{bmatrix} 3 & -10 \\ -2 & 7 \end{bmatrix}$

MATCHED EXAMPLE 8

$x = 1, y = 3$

MATCHED EXAMPLE 9

a $3a + 2b + c = 464$

$2a + b + c = 302$

$2a + 3b + 4c = 666$

b $\begin{bmatrix} 3 & 2 & 1 \\ 2 & 1 & 1 \\ 2 & 3 & 4 \end{bmatrix} \begin{bmatrix} a \\ b \\ c \end{bmatrix} = \begin{bmatrix} 464 \\ 302 \\ 666 \end{bmatrix}$

c It takes a machine 76 minutes to assemble model A, 86 minutes to assemble model B, and 64 minutes to assemble model C.

MATCHED EXAMPLE 10

a $\begin{matrix} \text{Month 1} \\ \text{Month 2} \end{matrix} \begin{bmatrix} 35 & 20 \\ 42 & 31 \end{bmatrix} \begin{bmatrix} 20 \\ 23 \end{bmatrix} = \begin{bmatrix} 1160 \\ 1553 \end{bmatrix}$

The total T-shirts' cost for month 1 is $1160.

The total T-shirts' cost for month 2 is $1553.

b $C = \begin{bmatrix} 1160 & 1553 \\ 3672 & 4250 \\ 4130 & 3280 \\ 3330 & 2050 \end{bmatrix}$

$S = \begin{bmatrix} 1392 & 1863.6 \\ 4406.4 & 5100 \\ 4956 & 3936 \\ 3996 & 2460 \end{bmatrix}$

c $P = \begin{bmatrix} 232.00 & 310.60 \\ 734.40 & 850.00 \\ 826.00 & 656.00 \\ 666.00 & 410.00 \end{bmatrix}$

Total profit: $4685

MATCHED EXAMPLE 11

a **i** Anna can send direct messages to Beth and David.
Beth can send direct messages to Anna and David.
Cho can send direct messages to Beth and David.
David can send direct messages to Beth and Cho.

ii The diagonal represents links where the sender and receiver are the same. This isn't considered communication, so they are redundant links.

iii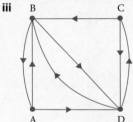

iv Cho → Beth → Anna

b

$$\text{sender} \begin{array}{c} \\ A \\ B \\ C \\ D \\ E \end{array} \begin{bmatrix} 0 & 0 & 0 & 0 & 1 \\ 0 & 0 & 0 & 0 & 1 \\ 0 & 1 & 0 & 0 & 1 \\ 1 & 0 & 0 & 0 & 0 \\ 1 & 1 & 0 & 1 & 0 \end{bmatrix} \begin{array}{c} \text{receiver} \\ A\ B\ C\ D\ E \end{array}$$

MATCHED EXAMPLE 12

a 1 **b** $D \to C \to A$

c 2 **d** $A \to C \to A$

MATCHED EXAMPLE 13

a **i** $x = 37\%$
$y = 44\%$
$z = 30\%$

ii This year

$\begin{array}{c} \\ A \\ B \\ C \end{array}$
$\begin{bmatrix} A & B & C \\ 0.4 & 0.44 & 0.15 \\ 0.23 & 0.2 & 0.55 \\ 0.37 & 0.36 & 0.3 \end{bmatrix} \begin{array}{c} A \\ B \\ C \end{array}$ Next year

b Current game

$T = \begin{bmatrix} W & L \\ 0.3 & 0.22 \\ 0.7 & 0.78 \end{bmatrix} \begin{array}{c} W \\ L \end{array}$ Next game

MATCHED EXAMPLE 14

a 1200 students **b** 660 students

c 2880 students **d** 15.6%

MATCHED EXAMPLE 15

a $\begin{bmatrix} 654 \\ 1526 \end{bmatrix}$

b In the long term, 654 ships will be at seaport R and 1526 ships will be at seaport S.

c 70%

CHAPTER 6

MATCHED EXAMPLE 1

a *Number of trees cut*

b *Age of a person*

c *Hours spent studying*

MATCHED EXAMPLE 2

a *age (years)*

b *number of pets*

c 14 people

d a 10-year-old who has 3 pets

e 4

f 2

MATCHED EXAMPLE 3

a **i** Positive, linear and strong

ii The weight of a person increases as their age increases.

b **i** Negative, linear and strong

ii The exam score increases as the time spent on watching TV decreases.

c **i** No association

ii There appears to be no association between the height of a person and the number of T-shirts owned by them.

d **i** Positive, linear and weak

ii There is limited evidence to suggest that the number of wristwatches owned by a person increases as the salary increases.

MATCHED EXAMPLE 4

a *Outdoor temperature* changes could be the underlying cause of the association between the two.

b *Battery capacity* could be the underlying cause of the association between the two.

c *Household income* could be the underlying cause of the association between the two.

MATCHED EXAMPLE 5

a	i	16.80	ii	17
b	i	40 444.00	ii	40 000
c	i	12.33	ii	13
d	i	0.17	ii	0.17
e	i	12 136.36	ii	12 000
f	i	6.53	ii	6.5
g	i	3.14	ii	3.1

MATCHED EXAMPLE 6

a intercept = 0.5

b *slope* = 0.13

c *time* = $0.5 + 0.13 \times age$

d 1.5 hours

interpolation

e 1.0 hours

extrapolation

f The prediction for the 8-year-old involves interpolation, so it is more reliable than the prediction for a 4-year-old, which involves extrapolation.

MATCHED EXAMPLE 7

a intercept = 890

This is the *wages* when the *experience* = 0.

This means that the weekly average wage for a fresher is $890.00

b *slope* = 8.8

This means on average the weekly wage of a person increases by $8.80 for every one-year increase in the *experience*.

CHAPTER 7

MATCHED EXAMPLE 1

a These two graphs are *not* isomorphic because they have different numbers of vertices and edges.

b These two graphs are isomorphic because they show exactly the same connections.

c These two graphs are *not* isomorphic because although they have the same number of vertices and edges, not all the connections are the same. In the first graph, *D* and *E* are connected, but in the second graph they aren't.

MATCHED EXAMPLE 2 ANSWERS

a i 6 vertices: *A, B, C, D, E, F*

   7 edges: *AB, AC, AD, BD, CD, DE, DF*

   3 faces: 2 enclosed and 1 outside the graph

   ii

Vertex	A	B	C	D	E	F	Sum
Degree	3	2	2	5	1	1	14

degree sum = 2 × number of edges

= 2 × 7

= 14

b i 2 7 vertices: *A, B, C, D, E, F, G*

   12 edges: *AB, AC, AD, AG BD, CD, CE, CF, DE, DF, EF, GG*

   7 faces: 6 enclosed and 1 outside the graph

   ii

Vertex	A	B	C	D	E	F	G	Sum
Degree	4	2	4	5	3	3	3	24

degree sum = 2 × number of edges = 2 × 12 = 24

MATCHED EXAMPLE 3

$$\begin{array}{c} \quad A\ B\ C\ D\ E\ F\ G \\ \begin{array}{c}A\\B\\C\\D\\E\\F\\G\end{array}\left[\begin{array}{ccccccc} 0 & 1 & 0 & 0 & 0 & 0 & 1 \\ 1 & 0 & 0 & 1 & 0 & 0 & 0 \\ 0 & 0 & 0 & 1 & 0 & 1 & 0 \\ 0 & 1 & 1 & 0 & 0 & 0 & 0 \\ 0 & 0 & 0 & 0 & 0 & 1 & 0 \\ 0 & 0 & 1 & 0 & 1 & 0 & 1 \\ 1 & 0 & 0 & 0 & 0 & 1 & 1 \end{array}\right] \end{array}$$

MATCHED EXAMPLE 4

a
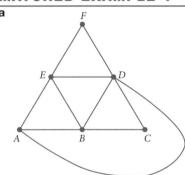

The graph can be redrawn without any edges crossing, so it is a planar graph.

b There is a path from each vertex to every other vertex, so it is a connected graph.

c *v* = 6, *f* = 6, *e* = 10

$v + f - e = 6 + 6 - 10$

$\qquad = 2$

Euler's formula works for this graph.

MATCHED EXAMPLE 5

a The number of vertices is 5.

b The number of edges is 19.

MATCHED EXAMPLE 6

a It is a subgraph because it only has vertices and edges from the larger graph.

b It is not a subgraph because it has a vertex (*H*) that is not in the larger graph.

c It is not a subgraph because it has an edge (*CD*) that is not in the larger graph.

d It is a subgraph because it only has vertices and edges from the larger graph.

MATCHED EXAMPLE 7

a circuit: no repeated edges, a repeated vertex, and starts and finishes at the same vertex

b walk only: repeated edge

c path: no repeated edges, no repeated vertices, and doesn't start and finish at the same vertex

d trail: no repeated edges, a repeated vertex, and doesn't start and finish at the same vertex

e cycle: no repeated edges, no repeated vertices (except the first and last vertex) and starts and finishes at the same vertex.

MATCHED EXAMPLE 8

a path: no repeated edges, no repeated vertices and doesn't start and finish at the same vertex

b cycle: no repeated edges, no repeated vertices (except the first and last vertex) and starts and finishes at the same vertex

c circuit: no repeated edges, a repeated vertex *V* and starts and finishes at the same vertex

d trail: no repeated edges, a repeated vertex *V* and doesn't start and finish at the same vertex

e walk only: repeated edge (*QV* is the same edge as *VQ*)

f cycle: no repeated edges, no repeated vertices (except the first and last vertex) and starts and finishes at the same vertex

MATCHED EXAMPLE 9

The shortest time needed to travel from city A to city B is 7 hours.

MATCHED EXAMPLE 10

a This graph is a spanning tree.

It has 8 vertices and 7 edges.

b This graph is not a spanning tree.

It has a cycle.

c This graph is a spanning tree.

It has 8 vertices and 7 edges.

d This graph is not a spanning tree.

It has a loop.

e This graph is not a spanning tree.

It has an edge (*AF*) that isn't in the original graph.

f This graph is a spanning tree.

It has 8 vertices and 7 edges.

MATCHED EXAMPLE 11

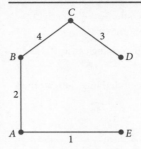

Total weight of spanning tree = 10

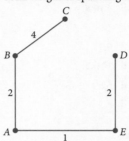

Total weight of spanning tree = 9

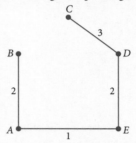

Total weight of spanning tree = 8

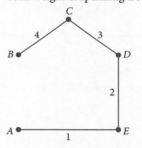

Total weight of spanning tree = 10

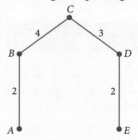

Total weight of spanning tree = 11

The total weight of the minimum spanning tree is 8.

MATCHED EXAMPLE 12

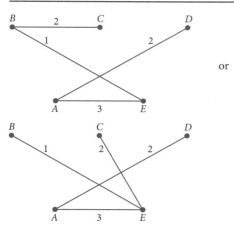

The total weight of the minimum spanning tree is 8.

CHAPTER 8

MATCHED EXAMPLE 1

a $c = km$

b $k = 1.96$

c

m (kg)	0.5	0.75	1.5	1.75
c ($)	0.98	1.47	2.94	3.43

d

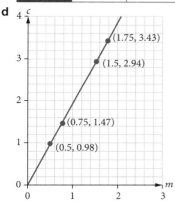

e 0.75 kg costs $1.47.

  1.5 kg costs $1.5 \times 1.96 = 2.94 = \$2.94$.

f The cost of 15 kg of wheat flour is $29.40.

MATCHED EXAMPLE 2

a $t = \dfrac{k}{n}$

b $k = 432$

c

n	8	16	24	36
t (days)	54	27	18	12

d

[Graph showing curve through points (8, 54), (16, 27), (24, 18), (36, 12) with axes t and n]

e Eight workers take 54 days to build the fence.

  Sixteen workers take $54 \div 2 = 27$ days to build the fence.

f It would take 6 days for 72 workers to build the fence.

MATCHED EXAMPLE 3

a The graph is a straight line that goes through (0, 0), so direct variation is involved.

  The equation of variation is $y = 2x$.

b The graph has the shape 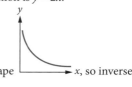 x, so inverse variation is involved.

  The equation of variation is $y = \dfrac{8}{x}$.

c The graph isn't a straight line and it doesn't have the

  shape 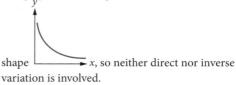 x, so neither direct nor inverse variation is involved.

d The graph is a straight line that doesn't go through (0, 0), so neither direct nor inverse variation is involved.

MATCHED EXAMPLE 4

a **i** The horizontal axis doesn't start at zero.

  ii Two points where one x value is double the other x value are (4, 8) and (8, 16).

  Double 4 is 8 and double 8 is 16.

  So, this is direct variation.

  iii The equation of variation is $y = 2x$.

b **i** The horizontal axis doesn't start at zero.

  ii Two points where one x value is double the other x value are (5, 8) and (10, 4).

  Double 5 is 10 and halving 8 is 4.

  So, this is inverse variation.

  iii The equation of variation is $y = \dfrac{40}{x}$.

MATCHED EXAMPLE 5

a **i**

x	0	1	2	3
y	0	3	12	27

ii

x	0	1	2	3
x^2	0	1	4	9
y	0	3	12	27

iii

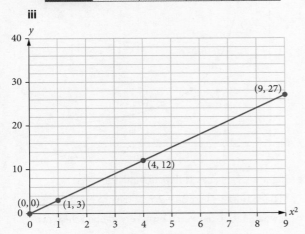

The x-axis label always matches the transformed variable.

b i

x	1	2	5	10
y	15	7.5	3	1.5

ii

x	1	2	5	10
$\frac{1}{x}$	1	$\frac{1}{2}=0.5$	$\frac{1}{5}=0.2$	$\frac{1}{10}=0.1$
y	15	7.5	3	1.5

iii

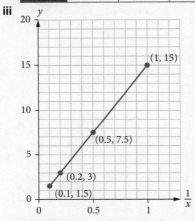

c i

x	1	2	4	5
y	0.6	0.9	1.2	1.3

ii

x	1	2	4	5
$\log x$	0	0.30	0.60	0.70
y	0.6	0.9	1.2	1.3

iii

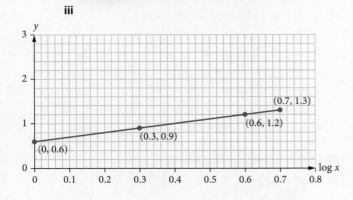

MATCHED EXAMPLE 6

a profit $= k\,(\text{cost})^2 + c$

b Fat % $= \dfrac{k}{(body\ mass\ index)} + c$

Fat % $= k \log (body\ mass\ index) + c$

MATCHED EXAMPLE 7

a i In two hours, the equation predicts the social networking app to have 62 new users.

  ii In five hours, the equation predicts the social networking app to have 79 new users.

  iii In ten hours, the equation predicts the social networking app to have 91 new users.

b

CHAPTER 9

MATCHED EXAMPLE 1

a $0.12\ \text{m}^2$ b $135\,000\ \text{mm}^3$
c $12.53\ \text{L}$ d $562\,300\ \text{mm}$

MATCHED EXAMPLE 2

a 4.52×10^{-6} b 1.226×10^7

MATCHED EXAMPLE 3

a 0.00235 b $624\,350$

MATCHED EXAMPLE 4

a i $x \approx 6.63\ \text{m}$ ii $x \approx 6.6\ \text{m}$
b i $x \approx 10.30\ \text{cm}$ ii $x \approx 10\ \text{cm}$
c i $x \approx 8.66\ \text{cm}$ ii $x \approx 8.7\ \text{cm}$

MATCHED EXAMPLE 5

$x = 5.3\ \text{cm}$

MATCHED EXAMPLE 6

The value of x is 1143 mm.

MATCHED EXAMPLE 7

A cyclist would travel 147 cm if she cycled directly up the centre of the ramp.

MATCHED EXAMPLE 8

a $AC \approx 7.94\ \text{cm}$ b $AD \approx 5.61\ \text{cm}$

MATCHED EXAMPLE 9

a i $P = 1600\ \text{cm}$ ii $A = 80\,000\ \text{cm}^2$
b i $P = 1160\ \text{cm}$ ii $A = 83\,200\ \text{cm}^2$
c i $C = 47\ \text{cm}$ ii $A = 177\ \text{cm}^2$
d i $P = 90\ \text{cm}$ ii $A = 390\ \text{cm}^2$

MATCHED EXAMPLE 10

a i arc length = 11.8 cm

   ii perimeter = 42 cm

   iii area = 177 cm^2

b i arc length = 50.3 m

   ii perimeter = 74.3 m

   iii area = 603 m^2

MATCHED EXAMPLE 11

a i 61 cm   ii 260 cm^2

b i 27 m   ii 45 m^2

MATCHED EXAMPLE 12

a 132 cm^2 b 33.6 m^2

MATCHED EXAMPLE 13

a The cost of fencing is $6332.

b The cost of mulch is $1878.

MATCHED EXAMPLE 14

a $V = 64$ cm^3

   $C = 64$ mL

b $V = 5540$ cm^3

   $C = 5.54$ L

c $V = 157$ m^3

   $C = 157\,000$ L

MATCHED EXAMPLE 15

a $V = 110$ m^3

   $C = 110\,000$ L

b $V = 570$ cm^3

   $C = 570$ mL

c $V = 96$ cm^3

   $C = 96$ mL

MATCHED EXAMPLE 16

a 1335 cm^3 b 6535 m^3

c 32 cm^3

MATCHED EXAMPLE 17

a 200 cm^2 b 39 m^2

MATCHED EXAMPLE 18

a 1300 cm^2 b 520 cm^2

c 110 cm^2

MATCHED EXAMPLE 19

a 7.5 m   b 39.1 cm

MATCHED EXAMPLE 20

a $\dfrac{12.4}{3.1} = 4, \dfrac{16.6}{8.3} = 2$

The scale factors are not the same, so the shapes aren't similar.

b $\dfrac{13.5}{4.5} = \dfrac{8.4}{2.8} = \dfrac{19.5}{6.5} = 3$

The scale factors are the same.

Use SSS.

The triangles are similar.

c $\dfrac{6}{4} = \dfrac{3}{2} = 1.5 \quad \dfrac{1.5}{3} = 0.5$

The scale factors are not the same, so the shapes aren't similar.

MATCHED EXAMPLE 21

a 4   b 1980 cm^2

c 9.5 cm^3

CHAPTER 10

MATCHED EXAMPLE 1

a $p \approx 19.46$ b $x \approx 4.94$

MATCHED EXAMPLE 2

The height of the ladder is 3.3 m.

MATCHED EXAMPLE 3

$\theta \approx 24°$

MATCHED EXAMPLE 4

The angle of inclination is 37°.

MATCHED EXAMPLE 5

a 26°   b 7 m

MATCHED EXAMPLE 6

a The three-figure bearing is 075°.

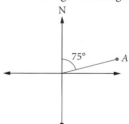

b The three-figure bearing is 125°.

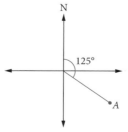

c The three-figure bearing is 315°.

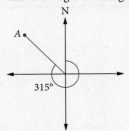

d The three-figure bearing is 232°.

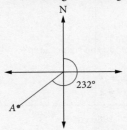

MATCHED EXAMPLE 7

a The barrage balloon is at a bearing of 326° from its starting point.

b The barrage balloon is 5 km from its starting point.

MATCHED EXAMPLE 8

a $x = 12$ cm   b $\theta = 26°$

MATCHED EXAMPLE 9

The height of the tree is 13.31 m.

MATCHED EXAMPLE 10

a $x = 29$ cm   b $\theta = 32°$

MATCHED EXAMPLE 11

a The distance that Rachel needs to travel to return to camp site X is 15 km.

b The three-figure bearing is 229°.